Table of Contents

Chapter 1.

Objectives and scope of this manual

Aims and objectives of the manual

Focusing restoration resources

Scope and layout of the manual

Why bother with reef rehabilitation?

Alasdair Edwards and Edgardo Gomez

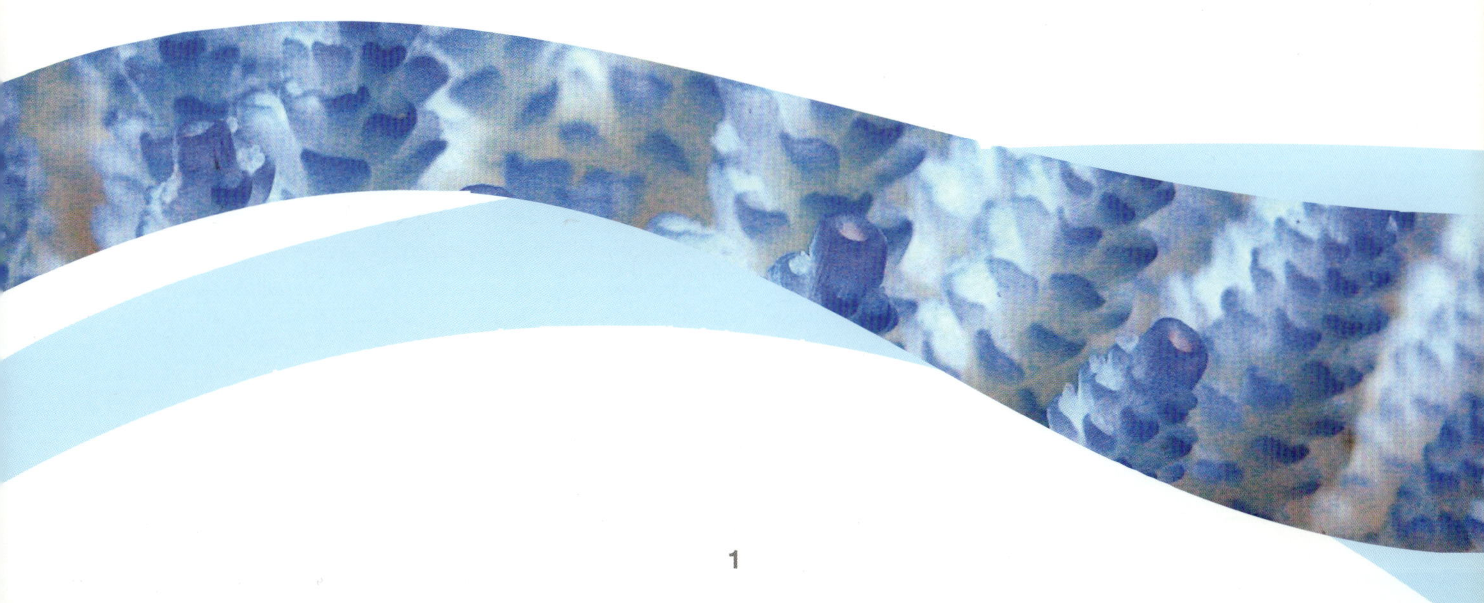

1.1 Aims of the manual

This *Reef Rehabilitation Manual* is intended to complement the *Reef Restoration Concepts & Guidelines*[1] and provide more detailed hands-on advice, based on lessons-learnt from previous experience, on how to carry out coral reef rehabilitation in a responsible and cost-effective manner. The two booklets should be used together. We build on the work of many people, notably Maragos (1974), Miller *et al.* (1993), Harriott and Fisk (1995), Heeger and Sotto (2000), Clark (2002), Job *et al.* (2003), Omori and Fujiwara (2004) and Precht (2006)[2-9], who have provided a considerable body of advice on restoring reefs (see References). Despite considerable advances over the last 35 years, coral reef restoration is still in its infancy as a discipline. A few rehabilitation projects appear to have been successful at scales of up to a few hectares; many, perhaps most, have failed or not met original expectations. The primary aims of this manual are 1) to reduce the proportion of reef rehabilitation projects that fail, 2) to introduce protocols for methods that could allow larger areas of degraded reef to be repopulated with corals whilst minimising collateral damage to reefs where corals are sourced, 3) to highlight factors to take into consideration at the planning stage so

as to minimise the risk of failure, and 4) to underline the current limitations of reef rehabilitation. The focus is on corals because these are the keystone species that give structure and topographic complexity to coral reef ecosystems. Unfortunately, they are also among the taxonomic groups most vulnerable to global climate change[10-11].

Bleached *Acropora* on the southern Great Barrier Reef in 2002 (O. Hoegh-Guldberg). Unfortunately, as a result of their symbiotic association with algae (zooxanthellae) and need to build calcium carbonate skeletons, reef-building corals are amongst those animals most vulnerable to global climate change.

We reiterate two important caveats[12]:

"Although restoration can enhance conservation efforts, restoration is always a poor[†] *second to the preservation of original habitats.*

The use of ex situ 'restoration' (mitigation) as an equal replacement for habitat and population destruction or degradation ('take') is at best often unsupported by hard evidence, and is at worst an irresponsible degradative force in its own right."

and one important definition[13]:

"Ecological restoration is the process of assisting the recovery of an ecosystem that has been degraded, damaged, or destroyed."

[†]For coral reef restoration this can also be a risky second choice (see Chapter 3).

1.2 Definitions of terms

It is perhaps useful also to consider definitions of restoration, rehabilitation, remediation and mitigation.

- *Restoration:* the act of bringing a degraded ecosystem back into, as nearly as possible, its original condition.

- *Rehabilitation:* the act of partially or, more rarely, fully replacing structural or functional characteristics of an ecosystem that have been diminished or lost, or the substitution of alternative qualities or characteristics than

those originally present with the proviso that they have more social, economic or ecological value than existed in the disturbed or degraded state.

- *Remediation:* the act or process of remedying or repairing damage to an ecosystem.

- *Mitigation:* the reduction or control of the adverse environmental effects of a project, including restitution for any damage to the environment through replacement, restoration, or creation of habitat in one area to compensate for loss in another.

Whatever the ultimate aims of management intervention (restoration, or the less ambitious goal of rehabilitation) or the reasons for the intervention (historical degradation, or as is often the case for mitigation, deliberately planned future degradation by a development project), the considerations and protocols that will maximise success are often the same. In all cases, what matters is that the management intervention is seeking to improve the state of the habitat and resilience of the ecosystem. Throughout this manual the terms "rehabilitation" and "restoration" will be used interchangeably, with in most cases the expectation being that active interventions are striving for some measure of "rehabilitation" with full "restoration" being unlikely in the face of global climate change impacts compounded by local anthropogenic pressures.

Many of the largest "reef restoration" projects are really compensatory mitigation exercises (often required by law), where reef habitat is being lost to development and parts of the ecosystem (mainly the corals) are moved to another site to make way for the development. This highlights how dangerous it is to overstate what restoration can achieve. If decision-makers believe that functioning reefs can be readily created by restoration interventions (e.g. transplanting reef organisms from a sacrificial site wanted for development to an area outside the impact zone, or by creation of artificial reefs), they will act accordingly. It should be emphasised to decision-makers that restoration science is still a long way from being able to recreate fully functional reef ecosystems and thus decisions which rely on compensatory mitigation are effectively promoting net reef loss. Further, compensatory restoration will only work if the same conditions and successional / disturbance history exists at the transplant site as at the source site. If this is the case, then similar coral communities should already be present at the transplant site. If a different coral community (or almost no coral) is present at a proposed recipient site, there is probably a good reason for this and efforts to introduce different coral species (that would change the structure and function of the local community) to such a site are likely to fail in the medium to long term. Mitigatory transplantation of corals should thus be seen as a last resort, and certainly not as a simple remedy.

There is a danger of equating the act of coral transplantation with "restoration" or "rehabilitation". The latter are long-term (decadal) processes; coral transplantation is an activity and just one tool among a suite of management measures that can be used to try to reverse a trajectory of reef decline. Although a few small-scale attempts at coral transplantation have delivered reasonable survival over a few years, this is not the same as restoration, just a first step towards a trajectory of improving ecosystem structure and function.

1.3 Focusing restoration resources

At a conservative estimate, there are around 255,000 km^2 of coral reefs worldwide in tropical seas[14]. An estimated 19% of these (c. 48,450 km^2) are considered severely degraded and a further 15% (c. 38,250 km^2) are thought to be under imminent risk from human pressures[15]. These areas will be a patchwork of reef habitats – reef pavement, algal ridges, massive coral stands, rubble, branching coral thickets, sand (maybe colonised by seagrass or rhizophytic green algae), and conglomerate – much of it consolidated but also much of it unconsolidated (e.g. sand and rubble). Given this, there must be many thousands of km^2 of consolidated reef substrate that is in need of rehabilitation. There is thus no shortage of hard substrate and a focus on trying to rehabilitate patches of consolidated reef that have lost their living coral would seem both a desirable aim for reef restoration and a relatively cost-effective approach whilst the science is still developing. Despite this, there has been inordinate attention paid to restoring or introducing coral cover to unstable rubble and sand patches with the concomitant need for (often costly) artificial structures ranging from iron bedframes to purpose built concrete and ceramic modules to which coral transplants can be attached. This is akin to carrying out the first reforestation experiments in areas without topsoil.

A macroalgae and *Drupella* infested *Acropora* stand close to the main township in the Funafuti lagoon (Tuvalu). Overfishing appeared to have reduced herbivore populations and nutrient inputs from the township were perhaps leading to eutrophication resulting in severely stressed corals with little chance of surviving any natural disturbance (D. Fisk).

It is useful to distinguish between *passive* restoration – that is, management actions that improve the environment, reduce overfishing, promote herbivory, etc., with the aim of promoting the natural recovery of reefs – and *active* restoration which involves direct interventions such as coral transplantation, removal of macroalgae (seaweeds), and substrate consolidation. Putting resources into implementing effective management is generally considerably cheaper than diverting them into active restoration measures such as coral transplantation. Further, at sites where there is significant local human impact on the reef, some form of management control (which will promote *passive* restoration) needs to be in place before any attempt at *active* restoration is made, otherwise the active interventions have a high risk of failure and consequently will be a waste of (often scarce) resources. Occasionally, there may be reef areas that are under effective management but do not appear to be recovering from a natural disturbance or previous human impacts (e.g. Ch. 8: Case study 1), where some active restoration intervention is needed in addition to good management in order to kick-start recovery[16].

Much *active* reef restoration has centred on ship-groundings because such events generate funds to repair damaged reefs. The scale of damage is typically in the order of 10^{-2} to 10^{-1} hectares (100 ha = 1 km^2) and, in a few instances, restoration attempts have met with some success at these scales. The largest attempt at active reef restoration has involved about 7 ha (or 0.07 km^2) of reef. These figures suggest that there is a six orders of magnitude mismatch between the area of degraded reef, and what *active* restoration can presently achieve. Furthermore, costs are substantial. If only active biological restoration (e.g. coral transplantation) is considered, then costs are in the order of several US$ 10,000's per hectare, not dissimilar to those for restoring other ecosystems such as mangroves, seagrasses or saltmarshes (see Chapter 8). But if damage to the reef framework is so severe that physical restoration (civil engineering) is needed, data from ship-groundings in the Caribbean suggest costs of US$ 2.0–6.5 million per hectare to repair injured reefs[17]. Thus trying to rehabilitate unconsolidated habitats tends to increase costs by 10 to 100 fold over those estimated for consolidated reef habitat.

1.4 Scope and layout of the manual

On the whole, coral reef rehabilitation is not a simple procedure that can be carried out by communities without training or expert advice. There are still many uncertainties and considerable care is needed at all stages if attempts at reef rehabilitation are to succeed. In addition, experience shows that bleaching events, coral predators and other unexpected disturbances are likely to hinder active restoration interventions, and that this is the <u>norm</u>, not the exception. Up to now, trained coral reef scientists have had variable success with experiments involving active restoration, so it is unrealistic and ultimately counterproductive to raise expectations that coral reefs can readily be rehabilitated. Reef restoration should never be oversold and its limitations clearly understood[18]. Having said that, techniques have improved greatly in recent years. Nursery rearing of corals from fragments or larvae allows a marked reduction in collateral damage to reefs (as a result of sourcing of transplants) and there have been big improvements in transplant survival and cost-effectiveness of methods.

At our current state of knowledge, we have some good ideas of what does <u>not</u> work, but still lack adequate experience to know what will work, particularly at a useful scale (several hectares). We are still learning what works and what doesn't work in a largely empirical way. We now know that tens of thousands of coral fragments can be

reared routinely into small colonies in *in-situ* coral nurseries, but we do not yet know whether these colonies can be deployed successfully over hectares of reef and generally survive to reproduce.

The following chapters seek to disseminate protocols that will, on the one hand, increase the chance of success of *active* restoration projects and on the other, reduce the impact of these projects on the natural reef if they fail.

For any large scale (hectares to km^2) reef rehabilitation, large numbers (tens to hundreds of thousands) of coral transplants are likely to be needed. To supply these, a two-step process is required. Firstly, small fragments of coral or coral spat need to be reared in nurseries to a size where they have reasonable prospects of survival on a degraded reef. Secondly, the nursery-reared colonies need to be transplanted to stable areas and attached securely. The three central technical chapters of this manual describe in detail how to construct and manage nurseries to farm coral fragments (Chapter 4), how to rear coral larvae for restoration (Chapter 5), and how to deploy coral transplants to degraded reef areas (Chapter 6). These chapters build on the previous work cited above and describe protocols which have been developed and tested in several countries during the GEF/World Bank's Coral Reef Targeted Research & Capacity Building for Management (CRTR) programme and the European Commission's REEFRES (*Developing ubiquitous practices for restoration of Indo-Pacific reefs*) project and over the last five years.

Coral reef in the Chagos Archipelago in the central Indian Ocean about one decade after almost all corals were killed to a depth of 10 m as a result of sea temperatures warming during the 1998 El Niño Southern Oscillation event. This illustrates the remarkable resilience of reefs that are undisturbed by local human impacts (N. Graham).

Chapters 2 and 3 seek to promote better use of the scarce resources available for reef rehabilitation by encouraging better project planning and management and explicit recognition of the risks inherent in *active* restoration approaches and ways of reducing these. The first of these chapters provides an overview of the steps needed in designing and planning a rehabilitation project. In particular, it examines criteria for deciding whether it is appropriate to attempt *active* restoration at a particular site (as opposed to implementing management measures that allow natural recovery), so that interventions can be focused where they a) have a reasonable chance of success, b) will make a difference in the long-term, and thus c) may be cost-effective. The second of these chapters recognises that corals are among the organisms that are most susceptible to climate change (e.g. rising sea temperature) and may be subject to large scale regional phenomena such as bleaching induced mortality and predation by Crown-of-thorns starfish (*Acanthaster planci*) over which rehabilitation project managers have no control. In addition, they may be threatened by local stressors which managers may be able to mitigate or control. By promoting a flexible rehabilitation plan with capacity for monitoring and adaptive management responses to changing needs, the impact of these and other risks may be mitigated to some extent.

Chapter 7 seeks to provide a comprehensive costing framework for rehabilitation projects. It looks at ways of costing reef rehabilitation projects so that others planning to carry out restoration can use the itemised costings to make realistic estimates of how much their restoration project may cost and judge what equipment, consumables and logistics may be required. Standardised and transparent costings are vital if valid comparisons are to be made between different restoration techniques and the cost-effectiveness of projects is to be evaluated. Once reliable costings are available then benefit-cost analysis (BCA) can be used to decide whether *active* restoration is an efficient allocation of resources at a location, or whether funds might be more cost-effectively spent on, for example, improving the enforcement of existing management regulations (*passive* restoration)[17,19].

Chapter 8 reviews lessons learnt from 10 case studies of coral reef rehabilitation projects from around the world. These have been selected to provide examples of the diversity of activities carried out under the broad umbrella of "reef rehabilitation". A summary of each case study is then presented in a standardised format with links to further information for each case-study for those who are interested.

Artisanal fishermen in Zanzibar who depend on coral reefs for their food and livelihoods (K. Kilfoyle).

We have tried to restrict the number of references cited to those which are the most pertinent for managers and others intending to undertake reef rehabilitation with a maximum of about 20 per chapter. Where possible, we have also sought out references that can be downloaded free-of-charge over the internet and provided URLs (web addresses). ReefBase (the official database of the Global Coral Reef Monitoring Network (GCRMN) and the International Coral Reef Action Network (ICRAN) at: www.reefbase.org) is a particularly good source of useful on-line references.

1.5 Why bother with reef rehabilitation?

Finally, given that global climate change is predicted to significantly degrade coral reef ecosystems within 50 years with a two-pronged onslaught of rising sea surface temperatures and ocean acidification[10-11], why even consider reef rehabilitation? Firstly, many coral reefs that are relatively free of human impacts have shown remarkable resilience to mass-bleaching and coral mortality such as occurred in 1998 in the Indo-Pacific. By contrast, those reefs that were already impacted by more localised human impacts such as overfishing or pollution have often shown little or no recovery. We thus infer that locally stressed reefs will have almost no chance of surviving the climate change impacts predicted for the 21st century (at least, in a form resembling what we consider a healthy "coral reef" today), whereas resilient ones will have a significantly better chance. Reef rehabilitation techniques are one tool of those trying to manage human impacts in reef areas. If these techniques, along with other local management interventions (fisheries regulations, MPAs, pollution control, etc.), improve ecosystem resilience, then those reefs have at least some chance of surviving as productive and functional systems (albeit with less biodiversity) in the face of the global impacts that cannot be managed at the local level[20]. Coral reefs currently provide food and livelihoods for hundreds of millions of coastal people in over 100 countries via the harvestable resources they generate[21], so anything that can contribute to their resilience and thus the food security of the peoples dependent on them seems a sensible use of resources.

References

1. Edwards, A.J. and Gomez, E.D. (2007) *Reef Restoration Concepts and Guidelines: making sensible management choices in the face of uncertainty.* Coral Reef Targeted Research & Capacity Building for Management Program, St Lucia, Australia. iv + 38 pp. ISBN 978-1-921317-00-2 [Download at: www.gefcoral.org/Portals/25/workgroups/rr_guidelines/rrg_fullguide.pdf]

2. Maragos, J.E. (1974). *Coral transplantation: a method to create, preserve and manage coral reefs.* Sea Grant Advisory Report UNIHI-SEAGRANT-AR-74-03, CORMAR-14, 30 pp. [Download at: nsgl.gso.uri.edu/hawau/hawaut74005.pdf]

3. Miller, S.L., McFall, G.B. and Hulbert, A.W. (1993) *Guidelines and recommendations for coral reef restoration in the Florida Keys National Marine Sanctuary.* National Undersea Research Center, University of North Carolina, at Wilmington. 38 pp.

4. Harriott, V.J. and Fisk, D.A. (1995) Accelerated regeneration of hard corals: a manual for coral reef users and managers. *Great Barrier Reef Marine Park Authority Technical Memorandum,* 16, 42 pp.
[Download at: www.gbrmpa.gov.au/corp_site/info_services/publications/tech_memorandums/tm016/index.html]

5. Heeger, T. and Sotto, F. (eds). (2000) *Coral Farming: A Tool for Reef Rehabilitation and Community Ecotourism.* German Ministry of Environment (BMU), German Technical Cooperation and Tropical Ecology program (GTZ-TÖB), Philippines. 94 pp.

6. Clark, S. (2002) Ch. 8. Coral reefs. Pp. 171-196 in Perrow, M.R. and Davy, A.J. (eds.) *Handbook of Ecological Restoration. Volume 2. Restoration in Practice.* Cambridge University Press, Cambridge. 599 pp.

7. Job, S., Schrimm, M. and Morancy, R. (2003) *Reef Restoration: Practical guide for management and decision-making.* Carex Environnement, Ministère de l'Écologie et du Développement Durable, IFRECOR. 32 pp.

8. Omori, M. and Fujiwara, S. (eds). (2004) *Manual for restoration and remediation of coral reefs.* Nature Conservation Bureau, Ministry of Environment, Japan. 84 pp.
[Download at: www.coremoc.go.jp/report/RSTR/RSTR2004a.pdf]

9. Precht, W.F. (ed.) (2006) *Coral Reef Restoration Handbook.* CRC Press, Boca Raton. 363 pp.

10. Hoegh-Guldberg, O., Mumby, P.J., Hooten, A.J., Steneck, R.S., Greenfield, P., Gomez, E., Harvell, D.R., Sale, P.F., Edwards, A.J., Caldeira, K., Knowlton, N., Eakin, C.M., Iglesias-Prieto, R., Muthinga, N., Bradbury, R.H., Dubi, A. and Hatziolos, M.E. (2007) Coral reefs under rapid climate change and ocean acidification. *Science,* 318, 1737-1742.

11. Reid, C., Marshall, J., Logan, D. and Kleine, D. (2009) *Coral Reefs and Climate Change. The Guide for Education and Awareness.* CoralWatch, University of Queensland, Brisbane. 256 pp.

12. Young, T.P. (2000) Restoration ecology and conservation biology. *Biological Conservation,* 92, 73-83.

13. Society for Ecological Restoration International Science & Policy Working Group. (2004) *The SER International Primer on Ecological Restoration.* www.ser.org & Tucson: Society for Ecological Restoration International. 13 pp.
[Download at: www.ser.org/content/ecological_restoration_primer.asp]

14. Spalding, M.D., Grenfell, A.M. (1997) New estimates of global and regional coral reef areas. *Coral Reefs,* 16, 225-230.

15. Wilkinson, C. (2008) *Status of Coral Reefs of the World: 2008.* Global Coral Reef Monitoring Network and Reef and Rainforest Centre, Townsville, Australia, 296 pp.
[Download at: www.reefbase.org/download/download.aspx?type=1&docid=13311]

16. Rinkevich, B. (2008) Management of coral reefs: We have gone wrong when neglecting active reef restoration. *Marine Pollution Bulletin* 56, 1821-1824.

17. Spurgeon, J. (1998) The socio-economic costs and benefits of coastal habitat rehabilitation and creation. *Marine Pollution Bulletin,* 37, 373-382.

18. Richmond, R.H. (2005) Ch. 23. Recovering populations and restoring ecosystems: restoration of coral reefs and related marine communities. Pp. 393-409 in Norse, E.A. and Crowder, L.B. (eds.) *Marine Conservation Biology: the Science of Maintaining the Sea's Biodiversity.* Island Press, Washington DC. 470 pp.

19. Spurgeon, J.P.G. and Lindahl, U. (2000) Economics of coral reef restoration. Pp. 125-136 in Cesar, H.S.J. (ed.) *Collected Essays on the Economics of Coral Reefs.* CORDIO, Sweden. ISBN 91-973959-0-0
[Download at: www.reefbase.org/download/download.aspx?type=10&docid=6307]

20. Hughes, T.P., Baird, A.H., Bellwood, D.R., Card, M., Connolly, S.R., Folke, C., Grosberg, R., Hoegh-Guldberg, O., Jackson, J.B.C., Kleypas, J., Lough, J.M., Marshall, P., Nyström, M., Palumbi, S.R., Pandolfi, J.M., Rosen, B. and Roughgarden, J. (2003) Climate change, human impacts, and the resilience of coral reefs. *Science,* 301, 929-933.
[Download at: www.sciencemag.org/cgi/reprint/301/5635/929.pdf]

21. Whittingham, E., Campbell, J. and Townsley, P. (2003) *Poverty and Reefs.* DFID-IMM-IOC/UNESCO. 260 pp.
[Download at: www.reefbase.org/download/download.aspx?type=10&docid=E0000006469_3]

A list of all references in the *Reef Rehabilitation Manual* with hyperlinks to free downloads, where these are available, is on the CRTR website at:
www.gefcoral.org/Targetedresearch/Restoration/Informationresources/RRM_Reference_links.htm

Chapter 2.

Steps in planning a rehabilitation project

Establishing the need

Initial scoping

Fact-finding for your rehabilitation plan

Developing a rehabilitation and monitoring plan

Alasdair Edwards and David Fisk

2.1 Introduction

Do you need to actively assist natural recovery?

Active reef restoration should be viewed as just one option within a broader integrated coastal management plan. It is not an alternative to management and unless the causes of reef degradation are under control, active restoration will ultimately fail. Whether active restoration is likely to be a cost-effective intervention depends primarily on 1) the causes of the degradation and 2) the state of the reef. The aims of restoration may vary considerably, from fisheries rehabilitation to restoration of benthic biodiversity to shoreline protection; all require different approaches. The main socio-economic reason to rehabilitate is to bring back the services (e.g., food security, shoreline protection) provided by healthy reefs to the hundreds of millions of people dependent on them.

The root causes of reef degradation can be split into those that can potentially be managed at a local scale and those that cannot. The former include a range of normally chronic, human-induced disturbances, such as sediment and nutrient run-off resulting from land-use changes, blast-fishing, coral mining, and overfishing. The latter include globally rising sea surface temperatures (SST), ocean acidification, El Niño-Southern Oscillation (ENSO) events (Box 2.1), tropical cyclones and tsunamis. Generally, healthy reefs – those relatively unaffected by chronic human impacts – appear to recover reasonably well from acute natural disturbances such as tropical cyclones and multi-year fluctuations in warm oceanic currents (e.g., ENSO) that cause mass-bleaching and mortality§. On the other hand, reefs that are already under chronic anthropogenic stress do not generally recover well from natural disturbance events.

A key factor in determining whether active restoration should

Satellite image showing Hurricane Mitch (an acute natural disturbance) sweeping through the Caribbean towards Belize and Honduras (NOAA).

be attempted is the current state of the local environment. At one extreme, if local environmental conditions are good, the degraded area is relatively small and there are no physical impediments to recovery (e.g., extensive loose rubble, lack of recruitment sources), a degraded patch of reef may recover naturally to a state that is more or less indistinguishable from its surroundings within 10 years. In such a case, active restoration may have very limited benefits. At the other extreme, if local environmental conditions are very poor (high nutrient inputs, sedimentation, overfishing, etc.) as a result of human impacts, the chances of re-establishing a sustainable coral population may be negligible. In such a case, major management initiatives (*passive* or indirect restoration) will be needed <u>before</u> any active restoration should be attempted. It is somewhat of an art deciding at what point along the continuum between these two extremes, active restoration is likely to be effective and what other management actions need to be taken before attempting restoration.

Need for underpinning management

It is important that effective management of an area is in place (unless there are no significant human impacts) before any attempts are made at active restoration. Active restoration <u>may</u> assist reef recovery once management is in place; it is almost certainly doomed to fail without effective coastal management. Therefore, active restoration should normally be limited to well-managed marine protected areas, sanctuaries, parks or areas under some form of *de facto* protection (e.g. resort reefs). However, if the area being managed is too small to mitigate the impact of all the key factors that are causing degradation, then those factors outside management control ("externalities") may prevent recovery. Thus, not only must there be effective management but it must be at a large enough spatial scale to allow those factors which caused the reef degradation in the first place to be controlled. Hence the need, stated at the start of this chapter, for active restoration to be considered as just one option within a broader integrated coastal management plan[1].

Rehabilitation of a habitat is always more expensive than protecting it from degradation in the first place, and the outcome is uncertain. Careful planning and implementation can reduce the risk of failure and improve the cost-effectiveness of the interventions. The rehabilitation project cycle can be split into five main stages (Figure 2.1). Decisions and activities at each stage need continually to be considered with respect to the financial and human resources available and the local social, economic and political environment. Of the five stages, only one involves implementation of rehabilitation interventions: the other four are there to try to ensure that the considerable investment in these is not wasted.

---

§ However, the frequency and intensity of these events could change with an increase in average SST from global warming, and this may have to be incorporated into the risk assessment of restoration viability.

Box 2.1 **Rising sea temperatures, ocean acidification and El Niño-Southern Oscillation (ENSO) warming events.**

So far, the most widespread acute damage to coral reefs has come from anomalous warming of the oceans associated with ENSO events. Sea temperature increases of 1–2°C above the long term maximum for a particular area are enough to trigger mass-bleaching of corals; if these temperatures persist for several weeks, this can be followed by mass-mortality as was seen in the Indo-Pacific in 1997–1998 and in the Caribbean in 2005. In the Indian Ocean in 1998, some areas experienced almost 90% mortality of shallow water corals. Many, but by no means all, such areas have shown good recovery within 10 years. In general, the less anthropogenically stressed reefs have appeared the most resilient.

These warming events are superimposed on a general trend of rising average sea surface temperatures of 0.1–0.2°C per decade which is inexorably pushing corals nearer their thermal stress limits. If mass-bleaching events become more frequent or more pronounced, then the outlook for coral reefs appears bleak as recovery processes may not have time to operate between events and coral cover will be continually ratcheted down. However, the scope for adaptation of corals and their symbiotic zooxanthellae remains unclear.

In addition to thermal stress the oceans are acidifying as burning of fossil fuels drives up the atmospheric partial pressure of CO_2, leading ultimately to predicted lower rates of calcification in marine organisms with calcium carbonate skeletons including corals, crustose coralline algae, molluscs and foraminiferans. Indeed, declines in coral skeletal growth and calcification over the last 30 years are already being reported. Model predictions suggest that as a result of this acidification, coral reefs could cease to grow and start to erode once atmospheric CO_2 doubles from pre-industrial levels to around 560 ppm, a level that is expected by 2100 even under the relatively conservative B1 scenario of the Intergovernmental Panel on Climate Change (IPCC), and earlier on less optimistic scenarios.

Thus coral reefs have a range of climate change related stresses to cope with over the coming decades in addition to local human impacts. If these local impacts can be ameliorated by better management then reef resilience can be improved locally. If local anthropogenic pressures continue, these are likely to result in an early loss of ecosystem function and services as climate change related factors impact them.

An Acropora *dominated reef undergoing mass-bleaching (O. Hoegh-Guldberg).*

Further information: For details of how to manage reefs in the face of mass coral bleaching and build long-term reef resilience, readers are referred to *A Reef Manager's Guide to Coral Bleaching*[1]. A review of possible futures for coral reefs under climate change[2] is summarised in a CRTR Advisory Paper entitled *Climate change: It's now of never to save coral reefs*[3]. The World Conservation Union (IUCN) paper on *Coral Reef Resilience and Resistance to Bleaching*[4] provides more detailed discussion of these issues for managers.

1. Marshall, P. and Schuttenberg, H. (2006) *A Reef Manager's Guide to Coral Bleaching.* Great Barrier Reef Marine Park Authority, Townsville, Australia. x + 163 pp.
[Download available at: coris.noaa.gov/activities/reef_managers_guide/reef_managers_guide.pdf or www.gbrmpa.gov.au/__data/assets/pdf_file/0015/13083/AReefManagersGuidetoCoralBleaching.pdf]

2. Hoegh-Guldberg, O., Mumby, P.J., Hooten, A.J., Steneck, R.S., Greenfield, P., Gomez, E., Harvell, D.R., Sale, P.F., Edwards, A.J., Caldeira, K., Knowlton, N., Eakin, C.M., Iglesias-Prieto, R., Muthiga, N., Bradbury, R.H., Dubi, A. and Hatziolos, M.E. (2007) Coral reefs under rapid climate change and ocean acidification. *Science*, 318, 1737-1742.

3. Coral Reef Targeted Research Programme (2008) Climate change: It's now or never to save coral reefs. *Advisory Paper*, 2 (1), 1-2.
[Download available from the Publications webpage on: www.gefcoral.org/]

4. Grimsditch, G.D. and Salm, R.V. (2006) *Coral Reef Resilience and Resistance to Bleaching.* IUCN, Gland, Switzerland. 52 pp.
[Download available (after running a search) at: www.iucn.org/resources/publications/publications_search/]

An area of reef reduced to rubble by blast-fishing. Note the colonisation by the soft coral Xenia (H. Fox).

Framework for a rehabilitation project

Before you do anything, you need to consider a series of initial questions (Figure 2.2) in order to decide whether active restoration might be a useful option. The aims of the rehabilitation also need to be agreed among all stakeholders. Secondly, if active restoration appears a sensible and useful option, you need to collect information on the site to be rehabilitated and on neighbouring reefs that may provide source material for transplantation so that the size of the task and its feasibility can be assessed. Thirdly, if active restoration appears feasible from the information collected, then you need to develop a detailed rehabilitation plan which can satisfy the aims agreed among stakeholders. This plan should be feasible in terms of local human and financial resources, should include an element of monitoring to allow progress to be evaluated, and should incorporate measurable and time-bound criteria for success, as well as feedback to stakeholders (Figure 2.1).

At any of these stages, you may discover that it would not be prudent to proceed and decide to abandon the rehabilitation attempt. Even once you are into the implementation stage, the built-in monitoring and evaluation may show that the project should be curtailed due to unforeseen circumstances (e.g., predator or disease outbreak, mass-bleaching event, or adverse social, political or economic changes). Chapters 3–6 focus on implementation and minimising the risk of poor ecological and socio-economic outcomes. This chapter focuses on the initial scoping, data collection and planning for a rehabilitation project using asexually produced coral fragments. The rationale will apply equally with the larval rearing techniques discussed in Chapter 5, however, these techniques are still largely experimental whereas the asexual techniques have been tried and tested by NGOs and communities collaborating with scientists.

2.2 Initial scoping [Stage 1]

To assist in the process of deciding whether active restoration should be attempted, a decision tree that addresses many of the key initial scoping questions, is shown in Figure 2.2. We look at these questions in more detail below. Firstly, you need to consider very seriously why active restoration is being chosen and convince yourself that passive restoration (e.g., other management measures) might not be just as effective (and much cheaper) in the long term. The series of questions (1.1.1 – 1.1.5 in Figure 2.2) is designed to focus your attention on the key factors which influence whether active restoration is a sensible option.

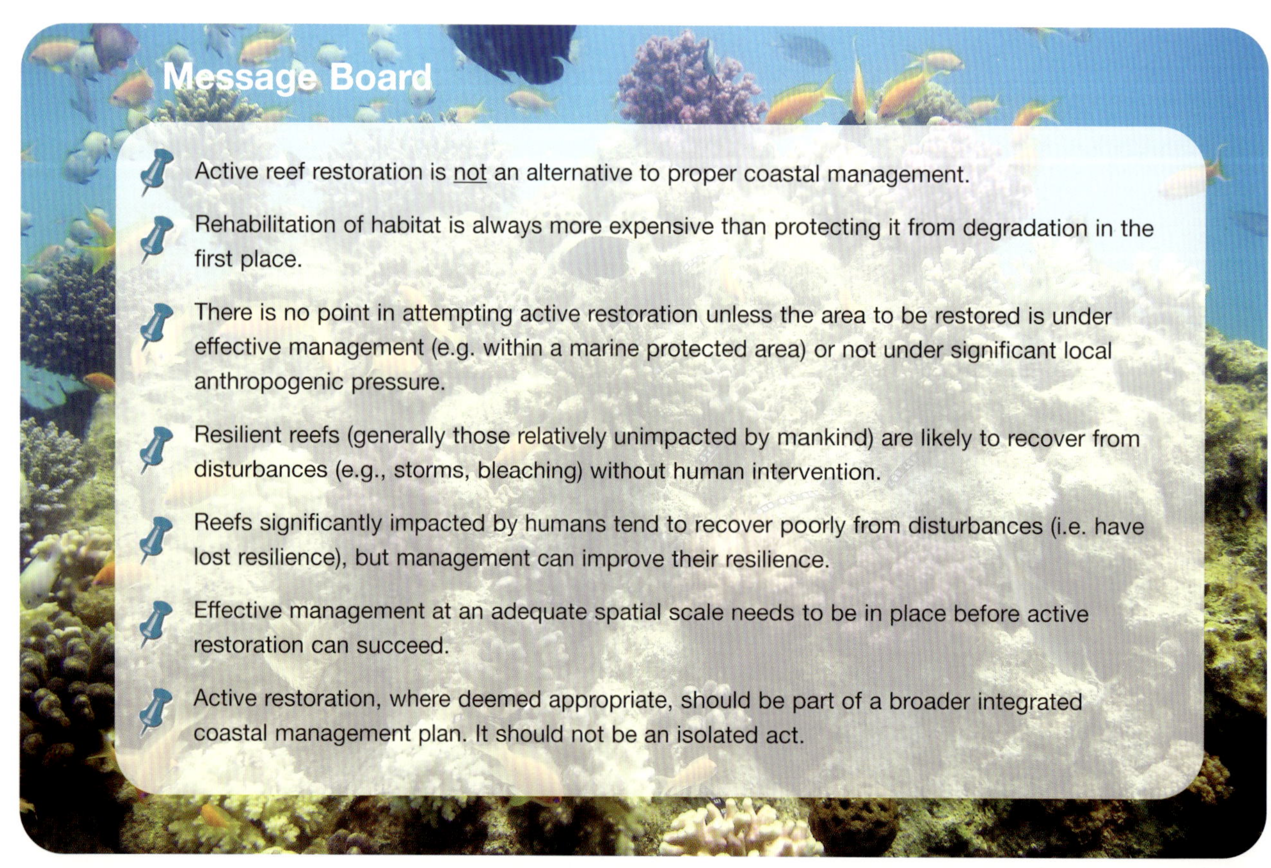

Figure 2.1. Five stages in a rehabilitation project. At all stages the proposed rehabilitation intervention needs to be considered in the context of local social and economic resources with a view to its sustainability. (See text for details.)

Did the site support a coral community prior to disturbance? [1.1.1]

For all "restoration" projects, this question should not need to be asked. How can you "restore" something that was not there before? However, for projects such as some tourism developments where there is a desire to create coral patches in safe sheltered sandy lagoon areas, this may be a pertinent question. This will also be the case in some mitigation projects where corals are to be transferred from sites that are being impacted by development to relatively bare sites nearby, or where corals are being transplanted to artificial reef structures. What type of coral community can survive at these sites? Ultimately, ecological constraints will determine this, not funding or human wishes. You need some knowledge of either what you are trying to restore or what coral community might be able to survive at the chosen sites. Bear in mind that even though a site may have supported a healthy and diverse coral reef community in the past, factors such as water quality may have

deteriorated and it may now only be able to support a few tolerant species.

What caused the degradation? [1.1.2]

Next you need to clarify what caused the degradation. This may be well-known, for example, if corals were badly affected by a single bleaching event or other memorable acute impact that is etched in local memory. On the other hand, if coral has disappeared slowly over 20–30 years due to multiple chronic human disturbances, the causes may be less clear and may be complex and diverse. If causes are unclear, then you should identify them as far as possible. If the causes are unknown, then you have a problem as you do not know if they persist or may re-occur and eventually kill your transplants. In such a situation it might be wise to consider a small pilot study to test how well transplants will survive. However, for this to be a useful gauge of local survivorship, it should be monitored over at least one full year and preferably longer.

Have the causes of degradation stopped? [1.1.3]

The aim of restoration is to restore a self-sustaining community. Therefore, once the causes of degradation have been identified, you need to look into whether the damaging impacts have stopped or are sufficiently under control so that they will not threaten any corals transplanted to the degraded site. For example, if water quality issues (e.g., nutrients from sewage, sediment from agricultural run-off, etc.) were part of the cause of degradation, until these have been dealt with by coastal management initiatives, there is no point in attempting active restoration. Beware that some impacts (e.g., sediment run-off) may be seasonal. If you explore issues fully with stakeholders, such transient impacts, which may not be apparent during short visits by expert advisors, are more likely to be identified. Obvious impacts are relatively easy to identify and site visits and discussions with stakeholders will indicate whether they have ceased or are under control. More subtle impacts are harder to identify and difficult to quantify. Such impacts can be caused by overfishing upsetting the balance between macro-algae (seaweeds) and corals.

If there is insufficient grazing (due to overfishing and/or loss of invertebrate grazers, such as sea urchins, through disease) and dense stands of macro-algae cover most hard substrate, then there is little chance of recruitment of corals (and other invertebrates) to establish the next generation. Transplants may survive for many years but if the ecological processes that allow them to produce future generations of young corals are compromised, the population is ultimately not sustainable. Without some management measures to restore ecological functioning, active restoration may be futile. At present we do no know what level of herbivory may be needed, but a survey can reveal whether there are many herbivores (e.g., parrotfish, surgeonfish, rabbitfish, urchins), the percentage cover of macro-algae, and whether there are any small corals (say < 5 cm) present. For example, if herbivores are rare, macro-algae are rampant and there's no sign of juvenile corals, this suggests that transplantation by itself will achieve little in the long term. Some other management measures (e.g., fisheries regulation, reduction of nutrient inputs) are needed first.

Often fleshy macro-algae, once well-established, are resistant to grazers with some species producing chemical compounds that make them distasteful to herbivores. In a few cases, preliminary data suggest that once management measures have allowed populations of fish and invertebrate herbivores to rebuild, these persistent seaweeds can be removed manually once and their return can then be kept in check by the herbivores. Thus algal dominated systems are not necessarily unredeemable.

Is the site recruitment limited? [1.1.4]

The next question is whether the site is "recruitment limited" (Box 2.2), that is, does it lack an adequate supply of coral

A terminal phase Stoplight parrotfish (Sparisoma viride) grazing fleshy algae and creating space for recruitment of corals and other invertebrates (R. Steneck).

larvae and are enough of these able to settle and survive?

On healthy reefs with a good natural supply of larvae and high post-settlement survival, there is likely to be little ecological need for active biological restoration. However, even on healthy reefs some areas will receive few coral and other invertebrate larvae in the currents and may recover more slowly from disturbances than those areas with a better larval supply. If recruitment appears very low, then the deciding factor is whether there are enough remnant corals that have survived disturbance and can serve as a base for recovery. If there is little sign of recruitment and very few remnant corals then using transplants to establish a viable local coral population may greatly accelerate recovery. After the 1998 mass-bleaching in Palau (Micronesia) it was found that the rate of recovery of coral cover at a series of sites little impacted by humans was highly dependent on the extent of remnant coral survival and not always correlated with larval recruitment rates[2].

Even if the cause of the coral loss has stopped, the site is not recruitment limited and natural recovery potential is high (such that it should recover unaided), there may be other reasons for active restoration, such as mitigation compliance, a political need for a restoration effort to be attempted (e.g. public outcry, concern, or insistence that an environmental injustice is corrected), or just human impatience with the rate of natural recovery. In such cases, given the large costs, the money made available for active restoration could probably be better spent on prevention of human impacts or on passive restoration measures (i.e. better coastal management).

Does the substrate require stabilisation? [1.1.5]

The final question relates to whether some physical restoration of the site is needed first to stabilise the substrate. If it is, this may be a very expensive precursor to transplantation efforts. If it cannot be afforded but is necessary, then attempts at active biological restoration are likely to fail. In such a situation, perhaps part of a site can be restored for the funding available.

Studies have shown that where blast fishing or coral mining

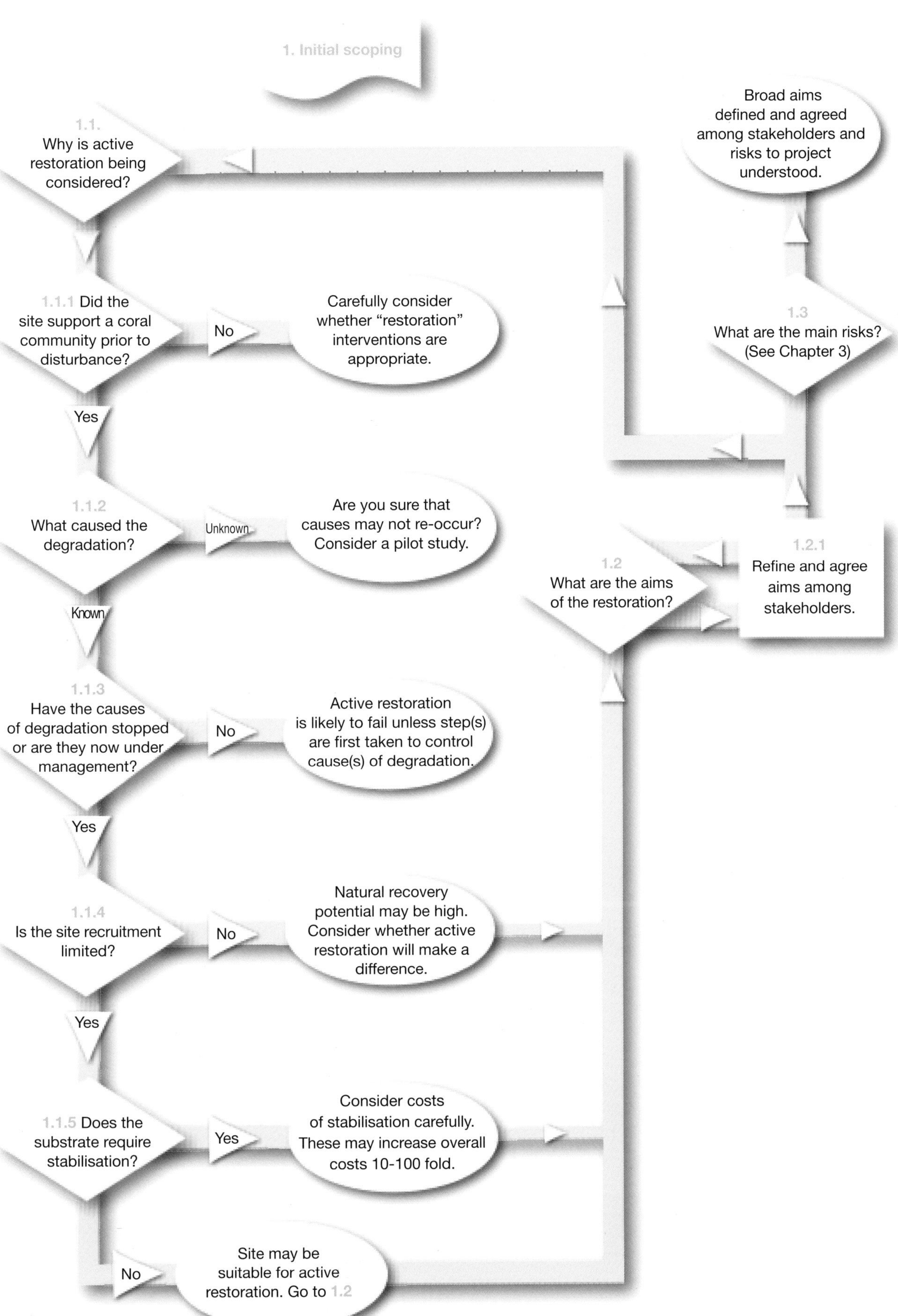

Figure 2.2. Decision tree for initial scoping questions to guide discussion of whether active restoration is an appropriate response to the reef degradation. If it appears likely to be useful, then the broad aims (expected outcomes) need to be discussed and agreed by stakeholders and the risks understood.

Box 2.2 What do we mean by "recruitment limited"?

By recruitment, we mean the addition of juvenile corals to the reef community. If there is good recruitment, you should find lots of small (0.5-5 cm) corals on the reef if you look carefully. For example, surfaces of bare ReefBall™ artificial reef structures had over 11 coral recruits per square metre within 10 months in an area with good larval supply in Palau. The average diameter was about 6 mm and the largest were almost 2 cm across. If you cannot find any, or only very few, juvenile corals, this suggests there is poor recruitment. If the numbers of juveniles recruiting are not sufficient to replace the adult colonies dying, the coral community is not sustainable and will ultimately decline.

Recruitment to a restoration site can be regarded as a three-stage process. Firstly, planktonic coral larvae have to reach the site. They can either be produced by spawning or planulation of remnant corals at the site (see Chapter 5 for details of coral reproduction) or be carried from nearby reefs with healthy coral populations by currents or tidal streams. If the restoration site has low connectivity with neighbouring reefs (which may be due to distance or current flows), recruitment will be dependent on local remnant corals that have survived the disturbance for larval supply. If it has high connectivity and currents are in the right direction, it may receive an abundant supply of coral larvae from neighbouring healthy reefs.

Left: *Plastic mesh being used to stabilise coral rubble, showing the successful colonisation by an* Acropora *coral recruit that, along with coralline red algae and other encrusting organisms, bind the rubble together (J. Maypa, 2004).*

Right: *10-month old recruits of* Favites halicora *settled and reared on an artificial substrate (a 'coral plug-in' consisting of a plastic wall plug with a 1.5 cm diameter concrete head) in the Philippines. Larvae were reared in tanks and many tens of coral spat initially settled on each plug-in but most died within a few weeks. After 10 months there are four surviving recruits with the largest polyp (about 5 mm in diameter) budding a new 'daughter polyp'. Note the dense growth of coralline red algae on the plug-in (J. Guest).*

Secondly, once coral larvae arrive at the restoration site, they have to find appropriate surfaces on which to settle and metamorphose into sessile (attached) polyps. A range of factors, including the topographic complexity of the site, presence of crustose coralline algae, amount of fleshy macro-algae, sediment build-up on surfaces, and amount of grazing by herbivorous fish and urchins to create bare space will determine what proportion of the larvae succeed in settling. In controlled conditions in tanks in laboratories, scientists have managed to get anything from 4% to 80% of larvae to settle (depending on the coral species). However, on a degraded reef subject to overfishing and other human pressures, it might be a tiny fraction of a percent that settle successfully.

Thirdly, the tiny polyp has to survive and grow for at least 6 months to a year until large enough for you to see it as a small "visible recruit" on the reef. Smothering by sediment, overgrowth by other sessile invertebrates (e.g., sponges, sea squirts) and algae, and predation will mean that only a tiny fraction of successfully settled polyps survive to this stage. Even with careful rearing of larvae in *in-situ* cages, only about 1% of settlers may survive to their first birthday. (Reducing these early huge losses at stages 2 and 3 in the recruitment process is the rationale behind larval rearing for restoration as described in Chapter 5.)

Thus a lack of observed recruitment may be due to different factors operating at different stages and requiring very different solutions.

Active restoration by coral transplantation is likely to be a useful intervention if the first stage (larval supply) is the recruitment limiting step. If the second or third stage processes (poor settlement, high post-settlement mortality) are the main issues, then passive restoration (management) measures to deal with overfishing or water quality issues will be needed first. In some instances, a combination of active (e.g., algal removal, coral transplantation) and passive (e.g., fisheries management) restoration might be the best solution.

have resulted in extensive areas of coral rubble, even if there is a good supply of coral larvae, these die soon after settlement where the rubble is unstable because of abrasion and smothering. Beds of coral rubble have thus been called "killing fields" for young corals. Coralline red algae, sponges and other encrusting organisms (including corals) can eventually stabilise rubble beds under certain conditions, but tropical cyclones and storms may interrupt the slow process of consolidation at any time. If the site chosen for rehabilitation necessarily includes areas of unconsolidated rubble, then you need to decide whether the rubble poses a major risk to coral transplants. In sheltered lagoonal environments, where the rubble is usually stable, you may decide that the risk is sufficiently low that you do not need to carry out stabilisation. On the other hand, where waves or currents are regularly moving the rubble, it must be stabilised in some way before there is any probability of recovery of the coral community. Indeed, where water quality and larval supply are good, substrate stabilisation may allow natural recovery without any need for transplantation (Ch. 8: Case study 1).

Various methods have been used to "stabilise" damaged reef areas. The difficulty and expense is proportional to the exposure of the site to waves and currents. At exposed sites, such as reef crests impacted by ship groundings, repairing physical damage is a specialist civil engineering task which can cost US$ 100,000–1,000,000's per hectare. For rubble fields exposed to moderately strong currents, islands of recovery have been created by deploying large limestone boulders on top of the rubble (Ch. 8: Case study 1) for costs of around US$ 50,000 per hectare and then letting natural recovery processes take over. Large quarried limestone boulders were also used successfully as part of the US$ 1,660,000 structural restoration following the M/V *Elpis* grounding in Florida in 1989 and survived subsequent hurricanes.

Limestone boulders used to rehabilitate a rubble field, showing results of several years of natural colonisation (H. Fox).

In the Philippines, patches of rubble created by blast-fishing have been successfully stabilised using 2-cm plastic mesh laid directly on the rubble and held down with rebar stakes[3]. These patches were at 8 m depth and relatively sheltered and the stabilisation resulted in about 10 times better survival of coral recruits than on adjacent unstabilised rubble leading to a significant increase in coral cover within two years. Rock piles made of reef rock and cement (0.5 m^2 in area and 1 m in height) were also placed on the mesh to attract fish and within two years the stabilised areas had fish communities similar to those on adjacent healthy reef in both species composition and biomass (Ch. 8: Case study 10). The estimated cost of this approach was US$ 44,000 per hectare if the whole rubble field were to be stabilised and US$ 13,750 ha^{-1} if rehabilitation islands were created (with a hundred 17.5 m^2 plots ha^{-1}). At a more exposed site in Maldives where sand scour and encroachment by mobile rubble were a problem, mesh was found not to be effective over a five year study. Thus, in some environmental settings, low-tech, relatively low-cost (US$ 10,000's per hectare) methods can be used to stabilise rubble areas, however, it is still too early to determine whether this approach will be successful from an aesthetic viewpoint. After 5 years, the plastic mesh remains conspicuous although part covered by sessile invertebrates and algae.

Coral reefs are mosaics of coral, sand, rubble, algal and seagrass habitats. If you can restrict your rehabilitation interventions to islands of stable substrate within the mosaic and your transplants will not be threatened by shifting rubble or sand in other patches, then leaving some areas unstable may be a more cost-effective option. Further, rubble patches can be an important habitat for many juvenile fish and invertebrates in healthy reef systems and are not necessarily symptoms of degradation.

What are the aims of the restoration? [1.2]

Working through these initial questions should assist you in answering the question "Why is active restoration being considered?" and clarify in your own mind, and those of stakeholders, why you think the site will not recover naturally with good management and why you think it is worthwhile investing effort and resources in assisting recovery by active intervention. You want the ecosystem services of the site to recover; the question is do you need to assist that recovery by active restoration or will it recover naturally once management measures are in place? Discussing these questions should also help the interested parties clarify what they want from the restoration and help build a consensus about the broad aims among stakeholders. For an active restoration project to be sustainable, the aims need to be articulated clearly and agreed among the stakeholders. Aims may differ among stakeholders but as long as they are not irreconcilable, complete agreement may not matter. A conservation NGO may be interested in improving biodiversity, whereas local fishers may be interested in

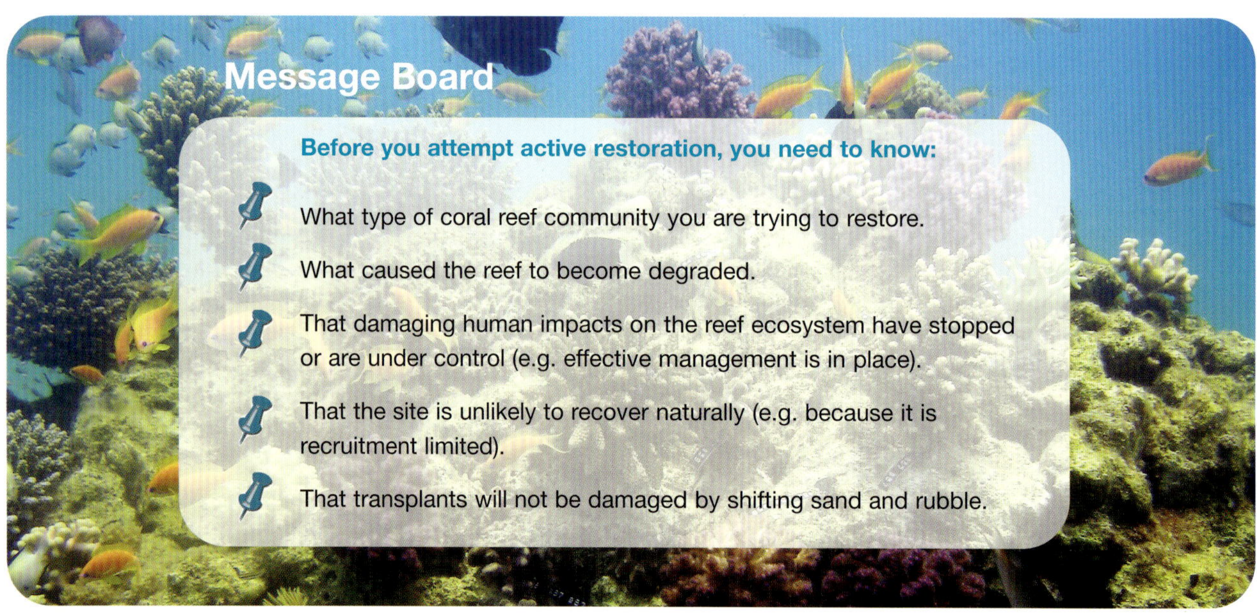

positive benefits in terms of improved fish catches. Both aims could be satisfied by a successful rehabilitation but will the proposed management structures be able to reconcile potential future conflicts between fishers wanting to access the recovering fish stocks and NGO members who want to protect the coral reef? Such issues are best considered early on in the project design process. The process of refining and agreeing aims among stakeholders may influence your thinking on why active restoration is being considered, hence the feedback loops in Figure 2.2.

The outcomes of working through the initial questions should be a decision as to whether or not active restoration is an appropriate response and, if it is, a consensus among stakeholders on the broad aims of the proposed restoration attempt.

What are the main risks? [1.3]

At this point it is worthwhile to consider the main (obvious) risks to the rehabilitation project (see Box 2.3 for a useful publication which can assist in this). Specific risks should be discussed more fully later in the planning process, during development of the detailed rehabilitation plan. Expectations

among stakeholders of achieving aims may be high. There is an old joke about the *Seven phases of a project* (see below). Following this, we may be at stage 2, having avoided stage 1 by our careful initial scoping. Risks and ways of trying to manage them are dealt with at length in Chapter 3. Here we just want to make the point that the expectations of the stakeholders will tend to be based on assumptions. These assumptions may be that the weather and sea conditions will be fairly normal during the project and that there will not be outbreaks of coral predators or disease. If these assumptions turn out to be wrong, then the project could be partly or wholly compromised, with few or none of the aims being achieved. If the risks are not discussed fully with stakeholders and nothing goes wrong, then no harm will be done. If the risks are not discussed and a Crown-of–thorns outbreak or bleaching episode destroys the corals at the rehabilitation site, then expectations are dashed and we move to stage 7 (dejected disillusionment). If the risks are discussed and recognised and stakeholders decide anyway to forge ahead with the project, then "it was just a risk". There may still be dejection, but not the disillusionment. The particular project may not have succeeded, but the whole concept of reef rehabilitation has not been discredited with the community.

Artisanal fisher in the Lingayen Gulf, northern Philippines (L. Raymundo).

Seven phases of a project

1. Uncritical acceptance.
2. Wild enthusiasm.
3. Escape of the clever.
4. Promotion of the non-participants.
5. Search for the guilty.
6. Punishment of the innocent.
7. Dejected disillusionment.

Box 2.3 *A Reef Manager's Guide to Coral Bleaching*

When considering 1) whether to attempt active restoration, 2) the likely risks in an era of rapid climate change, and 3) how your rehabilitation project might fit in with broader reef management in your area, a very useful free publication is: *A Reef Manager's Guide to Coral Bleaching* (Marshall and Schuttenberg, 2006).

This guide provides information on strategies that managers can implement as a short-term response to mass coral bleaching events and to support long-term reef resilience, with background information on the science and policies that support the management recommendations. Chapter headings are:

1. Managing for mass coral bleaching.
2. Responding to a mass coral bleaching event.
3. Building long-term reef resilience.
4. Coral bleaching – a review of the causes and consequences.
5. Enabling management – a policy review.

It is essential background reading for managers, NGOs and others wishing to become involved in reef rehabilitation.

The guide provides a conceptual context for attempts at active restoration and how these may fit in with management actions to promote recovery from disturbance. There are useful discussions of resistance, tolerance, resilience and factors influencing reef resilience, all issues of which it is important to be aware when planning rehabilitation projects. [See Box 2.1 for download details.]

2.3 Fact-finding for your rehabilitation plan [Stage 2]

If after working through the decision tree in Figure 2.2, you have decided that active restoration of the degraded site is likely to be appropriate and have agreed its broad aims, then you need to look into the <u>feasibility</u> of what you propose to do. Initially, this depends on local geography and ecology, human resources and financial resources. (Even if all these issues are positive then there are also likely to be social and political dimensions that you need to consider; these are likely to be unique to each project and there is no generic advice that can be given.) The first task is to quantify the scale of what needs to be done in order to initiate recovery and its feasibility on purely ecological considerations (Figure 2.3). This involves a field visit to collect and collate information about the site and neighbouring reefs. From this closer study, it may become apparent that although active restoration is desirable, it is not a feasible option in your local situation. On the other hand, if active restoration is deemed feasible, then once transplantation needs are quantified, the scale of the resources needed to progress towards the broad aims will become clearer. At this point you may identify a mismatch between your resources and aims and have the option of either seeking for additional resources or modifying your aims.

What areas within the site are suitable for restoration? [2.1]

The first step is to identify areas within the degraded site that are suitable for rehabilitation. (The overall suitability of the site has already been established in the initial scoping.) For a small ship-grounding, this might be the whole impact site; for a local community controlled marine sanctuary, it might be a number of denuded areas of coral rock within the sanctuary. The definition of "suitable" will partly depend on the broad aims and the human and financial resources available.

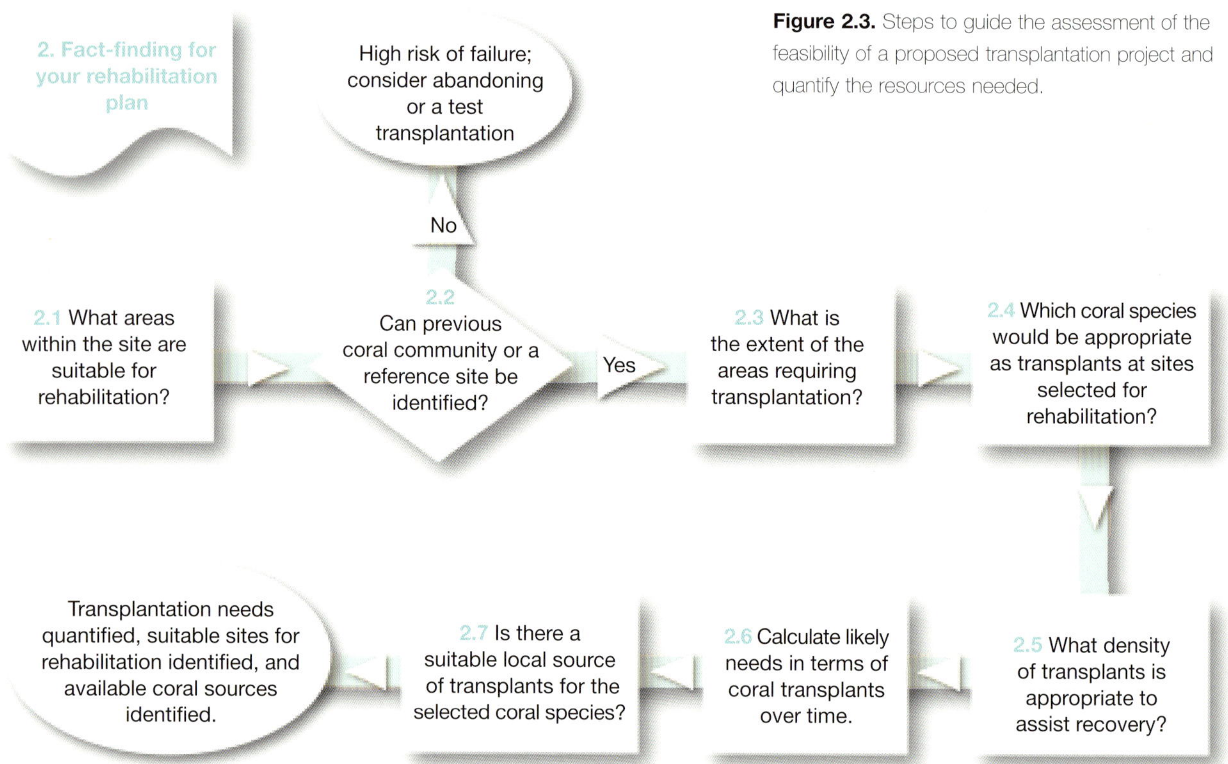

Figure 2.3. Steps to guide the assessment of the feasibility of a proposed transplantation project and quantify the resources needed.

Can previous coral community or a reference site be identified? [2.2]

Once the areas where active restoration is to be attempted have been identified, you need to decide what type of coral reef community you are intending to restore. If a healthy or remnant coral community survives in less damaged areas or on nearby reefs in a similar environment (depth, exposure, sediment regime, etc.), then this can serve as your "reference site". Alternatively, there may be records of what was there before degradation occurred (although this tends to be rare) and the growth forms of dead colonies in the degraded area can be indicative of what was present in the past. But beware that if local environmental conditions have changed significantly then species that were present historically, may no longer thrive. If you cannot identify what species might survive at the degraded site (based on the community at a reference site), then you have problems. Firstly, you run the risk that your transplants will be unsuited to the site and will die; secondly, if there is no reference site in the vicinity, then sourcing transplants is likely to be an issue as well. You should consider whether you might not be wise to either abandon the project or start with a small pilot transplantation to assess what species will survive well. However, note that even a pilot project will have to be run for at least a year to indicate species survival potential.

The "reference site" is a very important concept as it guides you in determining not only which species to transplant but also what a suitable density and mix of these species might be in the environmental setting you are trying to rehabilitate.

What is the extent of the areas requiring transplantation? [2.3]

Having identified suitable areas for rehabilitation within the site and a reference site, then you should measure the total area of the patches that you hope to transplant. The cost of transplantation will be proportional to the area. Are you attempting to restore tens, hundreds, thousands, tens of thousands (hectares) or hundred of thousands of square metres of reef?

Which coral species would be appropriate as transplants at sites selected for transplantation? [2.4]

The reference site is used to decide what common coral species ought to survive at the transplantation site. Although it provides a long-term goal for your restoration, in the short-term you are not trying to replicate the reference site but using it as a guide as to which species to transplant to assist natural recovery and their relative abundance. Fast-growing branching species (acroporids and pocilloporids) can act as "engineering species" as they can quickly generate topographic complexity and provide shelter for small fishes and invertebrates. Massive and sub-massive framework builders such as poritids and faviids tend to be slower growing but tend also to be less susceptible to bleaching, disease and predators. To minimise the risks (Chapter 3), you should attempt to transplant a broad cross-section of species with due regard to both the reference site and the availability of source material [2.7].

What density of transplants is appropriate to assist recovery? [2.5]

Some estimate of the likely density of transplants that you hope to achieve is now needed. The reference site can provide you with an idea of the density of coral colonies (of proposed transplant size or bigger) on a healthy reef of the type you are aspiring to. This gives you an estimate of how many transplants per square metre you might aim to deploy. For reasons of cost-effectiveness you will likely wish to transplant at the lowest density that will be effective. If you double the distance between transplants, you will need four times less transplants and quarter the costs of transplantation. This power relationship means that the decision on density is a very important one.

There is debate as to whether it is most cost-effective to deploy transplants uniformly over a site or concentrate transplants to create islands of recovery within the degraded site. As yet, science cannot provide any clear answers but there are a range of issues that can be considered. If a uniform density turns out to be unsuccessful, then the project has gambled on that density and lost. The options are then remedial management with further transplantations at other densities (at considerable expense) or abandonment. On the other hand, if a series of patches are rehabilitated at varying (two or more) densities within a reefscape, the chance of all failing should be less. Small spatial scale transplantations in areas otherwise poor in corals have proved vulnerable to attacks by predators such as Crown-of-thorns (*Acanthaster planci*), cushion stars (*Culcita*) and corallivorous snails (e.g. *Drupella*). Spacing transplanted patches across the rehabilitation site may allow predator outbreaks to be noticed (and responded to) in one patch before all of the transplantation has been decimated. Gaps between transplanted plots can act a bit like fire-breaks in forests as long as routine monitoring is being carried out sufficiently frequently to notice predator

outbreaks before the whole site has been ravaged. At some sites, only part of the site may have been identified as suitable for transplants anyway [2.3], militating patchiness. Your mindset with regard to reef restoration will influence your choice of density. If your mindset is that you wish to establish the target community of the reference site in a single, large, human intervention, you will clearly transplant at a density close to that at the reference site (or higher if you anticipate significant mortality) and cover as much of the site as feasible. If your mindset is that transplantation is just one step in an incremental process (over many years) to assist the target coral community to become established in the long term, you will be seeking to transplant at the lowest density that can kick-start recovery of patches within the site. The expectation in this case would be that these patches will expand over time through growth and export of fragments and sexual recruits, and through providing shelter for herbivores that graze algae and allow better local recruitment, etc. If your expectations are not fulfilled then you should have plans for adaptive management (e.g., transplanting additional patches or infilling existing patches) depending on monitoring results. At our current state of scientific knowledge, the first mindset is essentially a gamble. It may be successful, but experience suggests it is more likely to fail. The second mindset which seeks to work with "nature" involves a longer term outlook and allows learning from the progress (or set-backs) of the rehabilitation project; it is also likely to be more cost-effective. In addition, it encourages a sense of community stewardship of the reef and ownership of the project.

When estimating the density, think what the transplantation site might look like in one, two or more years after transplantation given various survival rates and growth rates and the initial size of transplants.

For the rehabilitation to be sustainable, natural recovery processes of growth, reproduction and recruitment need to be functioning. These processes should be augmenting the effects of the active restoration and should be factored in when deciding on densities. Because natural recovery tends to take at least 10 years, then your timeframe should be similar with final goals 5–10 years or more into the future, but intermediate milestones at appropriate intervals (e.g., annual) along the way.

Calculate likely needs in terms of coral transplants over time [2.6]

Once you have estimated the extent of the area you wish eventually to rehabilitate and have an idea of the density of transplants you would like to deploy, you can calculate the likely initial needs in terms of transplants. At this early stage, where many decisions still remain to be taken, it is sensible to provide estimates for a range of scenarios. A hypothetical example is given below (Box 2.4) to illustrate how density and patchiness decisions can be crucial in terms of logistics.

Crown-of-thorns starfish devastating an Acropora *dominated reef at Bolinao, Philippines in 2007 (K. Vicentuan).*

Box 2.4 Reality checkpoint

With 19% of the world's reefs severely degraded and a further 15% under serious threat over the next 10–20 years (*Status of the Coral Reefs of the World: 2008*[4]), there must be several 10,000's km² of reef in need of rehabilitation (mostly through improved management). There is much talk of "large-scale" restoration but the reality is small-scale, mostly sub-hectare. The largest project to date appears to have restored about 7 ha. The hypothetical example below looks at an area of 3 ha (or 0.03 km²) where transplants need to be affixed to the substrate. It is not large-scale by reef degradation standards but is quite large-scale by reef rehabilitation project standards.

Studies where individual colonies have been transplanted and fixed to degraded reef with epoxy putty suggest that about 4–5 colonies are normally transplanted per hour per person involved. The table below looks at time-needed in person-days to transplant corals to a 3 ha site assuming that 6 or 12 colonies per hour are being attached per person under three different scenarios. The rates assume that the current method has scope for improvement with practice.

Total area within rehabilitation site suitable for transplantation = 3 ha Density of colonies > 10 cm diameter at reference site = 8/m²	Transplant density (#per m²)	Number of transplants required	Person-days @ 6/h	Person-days @12/h
Scenario 1: transplant all suitable areas (3 ha) in one intervention in year 1	8	240,000	5,000	2,500
Scenario 2: transplant one third of site (1 ha) in patches in year 1	4	40,000	833	416
Scenario 3: transplant one third of site (1 ha) in patches in year 1 at 3 different densities (each over 0.33 ha). (If monitoring suggests need, then additional transplantation may be considered later.)	8 6 4 Total	26,667 20,000 13,333 60,000	556 416 278 1,250	278 208 139 625

Assuming that one transplant can be fixed to the reef every 5 minutes, for scenario 1, the results suggest that 10 people working for about one year might be able to carry out the transplantation task, whereas for scenario 3, the task could be achieved by five people working for about 6 months. Such calculations may surprise you and may even force re-evaluation of your aims as you work through the logistics. Other scenarios such as incremental transplanting over several years can be explored in a similar way. For community-based rehabilitation projects with limited resources, incremental designs may be the best approach. These are likely to suit local communities who are there for the long-term, but may not suit managers looking for quick fixes. The main point of this hypothetical (but realistic) example is that even interventions over relative modest areas require a significant investment in time and resources.

Is there a suitable local source of transplants for the selected coral species? [2.7]

Having identified the species [2.4] and estimated the numbers [2.6] of transplants needed, the next step is to locate potential sources of transplant material that are near enough the rehabilitation site to allow the corals to be transported in good health. Experience suggests that the source site(s) should be no more than 30–60 minutes away by boat unless special facilities are available to hold the corals during transportation. The transplants could be derived from "corals of opportunity" (natural fragments on the reef that have a poor chance of survival) or could be fragments removed from intact colonies at the reference site or similar reef. You need to estimate how much source material can be readily obtained without causing significant damage to the donor areas.

Translocation of large corals in a submerged cage towed by a boat during a mitigation project in Mayotte (Carex Environnement).

Studies suggest that 500 small fragments of a size suitable for *in-situ* nursery rearing can be obtained from a single, 20 cm diameter, branching coral colony. Thus 10% of a donor colony (maximum amount that it is recommended to remove) might yield 50 fragments. A recent study suggests that 35 randomly sampled donor colonies (widely spaced to avoid sampling clones) will retain a major proportion of the original genetic diversity of a population[5], thus making it feasible to preserve adaptive variation and to avoid problems such as inbreeding depression[6]. To estimate the potential supply of corals of opportunity, you could lay a plastic measuring tape (or rope marked in metre increments) over a potential source-site reef and count all corals of opportunity of selected species which lie within 0.5 m (or 1 m) either side of the tape. A study of five areas in a degraded lagoon in Philippines indicated an average of about 1–7 detached fragments per square metre of reef with average geometric mean diameters ranging from 2.4–5.3 cm. About 10 species were represented in a sample of 620 fragments. This snapshot suggests yields of tens of thousands of corals of opportunity per hectare may not be unusual.

If there is no suitable local source of transplants (i.e. within a few hours travelling distance) for the selected coral species, then you have a logistical problem. The magnitude of the problem will be proportional to the distance that corals will need to be transported. As part of the global trade in marine ornamentals, live coral is transported between continents by air with over a million pieces traded in some years[7]. However, packing and conditions are very carefully controlled to allow survival. Thus the technology is there, but at a cost. Seek expert advice if you need to transport for more than an hour or so and follow the coral collection recommendations in Chapter 4 (e.g. keep genotypes separated). At all times ensure the seawater in which the corals are held is kept well-aerated and try to keep the temperature within 1–2°C of that at the collection site to minimise stress.

Summary

The outputs from the fact-finding mission are:
1) a much clearer idea of the magnitude of the undertaking,
2) knowledge of the location and extent of areas within the site that are suitable for rehabilitation,
3) estimates of how many transplants might be needed, and
4) identification of a site or sites from where they can be sourced.

The feasibility or otherwise of the project will be much clearer and at this point you should again consider your findings with respect to likely available human and financial resources. If the project still appears feasible, then with all these facts and figures collated you are now in a position to develop a rehabilitation plan. A lot of very useful advice which can help you with project design and implementation is available free over the internet (e.g. Box 2.5) and should

be used to complement the recommendations made here.

2.4 Developing a rehabilitation plan [Stage 3]

Armed with quantitative information from your fact-finding, the next stage is to use this to develop a feasible plan for the rehabilitation with due consideration of the human and financial resources available. If progress towards the agreed aims of the project is to be assessed, then this plan must include both measurable and time-bound objectives and some element of monitoring [Stage 5] so that stakeholders can determine whether the objectives have been achieved. Interim objectives and monitoring are also necessary if any adaptive management of the project is proposed.

Is nursery rearing required? [3.1]

The first decision to make is whether nursery rearing (see Chapter 4 for detailed discussion of the rationale for it) is needed to generate enough transplants. In a mitigation project where an area of reef is being sacrificed to development of some kind (e.g., port development, pipeline construction) there may be large amounts of coral material that are being "rescued" and no need for more source coral. In such a case, a nursery stage may not be needed although it could be used imaginatively to multiply the rescued material. However, for any projects aiming to rehabilitate denuded areas of a hectare (100 m x 100 m) or more then tens of thousands of transplants are likely needed and a nursery rearing step is advised to reduce the collateral damage to donor reefs and in some cases may be the only realistic way of generating enough transplant material. If nursery rearing is indicated, then use Chapter 4 to decide what type of nursery might work and identify a sheltered site in the vicinity where a nursery can be safely constructed and operated. For small-scale transplantations (tens to a few thousands of m²) it may be possible to work with direct transplantation using fragments from donor colonies and/or "corals of opportunity", with, if necessary, some of the first generation of transplants being used as sources of further material after a period of growth.

Diver maintaining corals in a mid-water floating coral nursery in the Red Sea (S. Shafir).

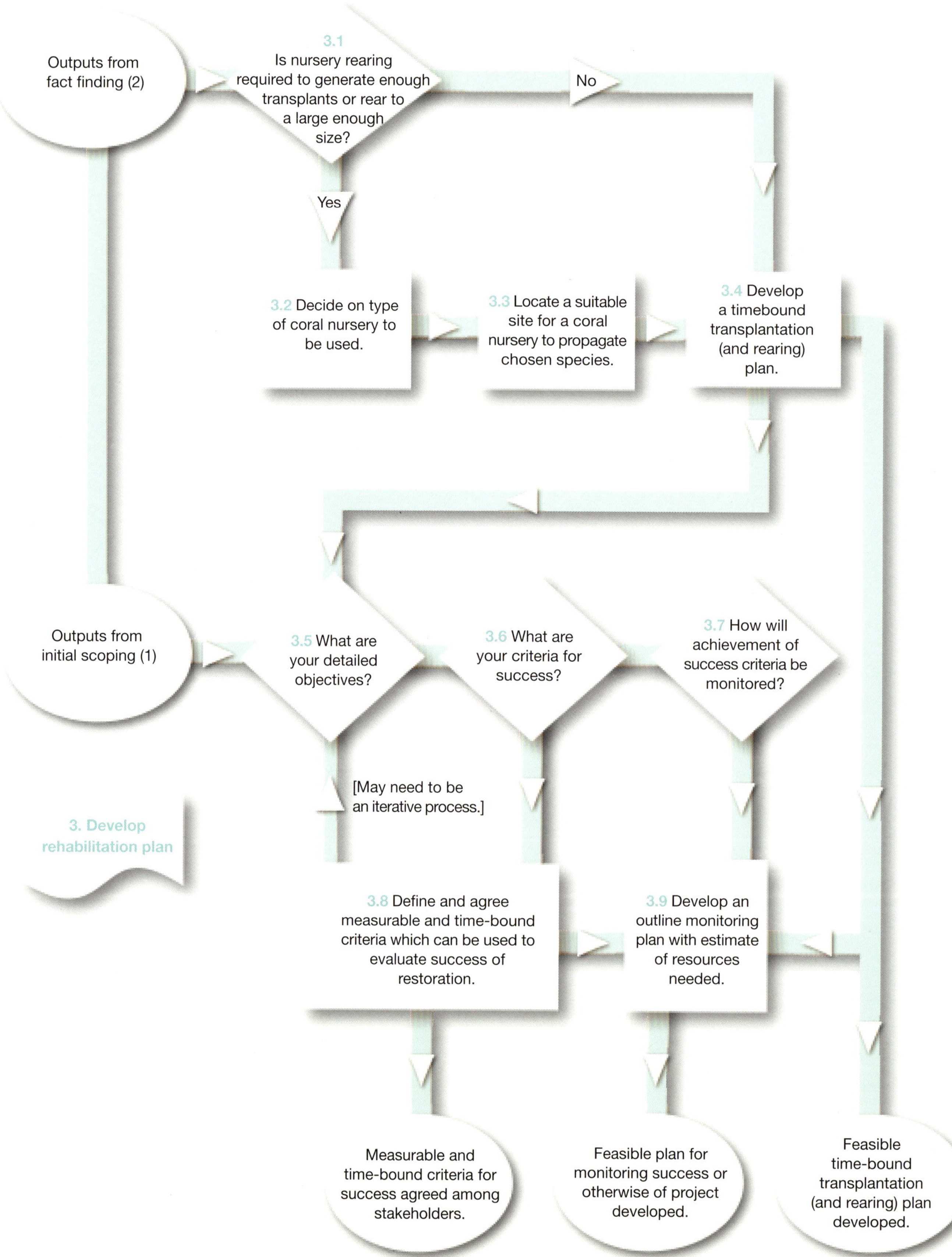

Figure 2.4. Steps and decisions in developing a detailed rehabilitation plan.

A key decision in developing a time-bound plan is whether to go for a staged (incremental) approach with repeated rearing cycles and transplantations with each cycle learning from the results of previous ones, or a bolder and more ecologically risky larger scale approach, with major early effort but less scope for adaptive management as the project progresses.The staged approach has the advantage that total resources needed at any one time can be quite modest with new areas being transplanted each year until the aims are achieved. It has the disadvantage of requiring longer commitment to the project which may increase the social and political risks (e.g. loss of stakeholder or local political support). Chapters 4 and 6 provide advice on rearing corals in nurseries and transplantation and should guide you in developing a plan suitable for your local situation.

What are your detailed objectives? [3.5]

Without clear objectives, it is not possible to evaluate success and it is difficult to learn lessons. With a detailed and time bound rehabilitation plan you are now in a position to refine the broad aims agreed among stakeholders into a series of more precise objectives that you hope to achieve at various times after the start of the project. These could be a series of annual milestones (or targets) which you and the other stakeholders think ought to be achievable if the project is succeeding. The objectives need to be realistic and "objectively verifiable" (i.e. their achievement or otherwise can be evaluated from the results of the planned monitoring§). The indicators should match the aims/objectives so that, if the targets are attained, then the aims/objectives will have been successfully achieved. Thus a time-bound verifiable objective of "transplanting 10,000 corals by the end of year 1" is inappropriate as it does not measure progress towards restoration – it could be achieved and there could be 10,000 dead transplants on the reef covered in filamentous algae. On the other hand, an objective of "increasing live coral cover by 20% by the end of year 1" is better.

An explicit timeframe with milestones allows the progress of the restoration to be monitored over time and corrective actions (adaptive management) to be undertaken if appropriate, such as when indicators fail to perform within the predicted timeframe. Indicators may be endpoints such as percentage live coral cover or evidence of restoration of key ecosystem processes such as coral recruitment or fish grazing.

If achieving a similar state in terms of coral cover and fish community to a healthy reference site over a defined period was part of the broad aims, then positive changes in these two indicators over time would be useful criteria for judging progress. Given uncertainties in rates of recovery, early milestones should concentrate on direction of changes in indicators rather than absolute levels.

What are your criteria for success? [3.6]

Whilst refining your objectives you should also consider defining your criteria for success and how you will assess whether these are achieved. This is seldom done in reef restoration projects and allows projects which ultimately fail to be passed off as achieving some "success" on the basis that some corals were transplanted and some were still alive after several months. An honest statement of stakeholders' aspirations (broad aims), followed by clear objectives with measurable criteria that will demonstrate their achievement allows both stakeholders and others to judge the rehabilitation project. Obscuring the issue neither benefits the reefs nor the communities that depend on them and is likely to be counterproductive as communities become disenchanted and disillusioned by oversold claims that cannot be delivered. The very process of discussing aims, refining objectives and seeking stakeholders' concepts of success criteria will help to avoid misapprehensions and lead to realistic and achievable objectives, so avoiding later disappointment. Specific risks to the project from external factors outside your control (Chapter 3) should be explained more fully to stakeholders at this stage.

The main output of this process are a defined and agreed set of measurable and time-bound criteria [3.8] which can be used to evaluate the progress of the restoration with respect to the objectives. Such a criterion might be: "Fish biomass at the rehabilitation site should increase by 10% within one year" if one of the aims of the project was to increase reef fish stocks.

How will achievement of success criteria be monitored? [3.7]

In Figure 2.1, stage 5 (Monitoring, evaluation and feedback to stakeholders) was separated from stage 4 (Implement rehabilitation plan) for clarity. In reality, these two stages should be integrated (as in Figures 2.4 and 2.5) as monitoring forms a necessary part of the implementation. For scientific research projects on coral transplantation or reef recovery, considerable time-consuming monitoring may be carried out using methods detailed in such publications as *Methods for Ecological Monitoring of Coral Reefs: A Resource for Managers*[8] or *Survey Manual for Tropical Marine Resources*[9]. Much of the detailed monitoring done by scientists may be irrelevant in terms of assessing progress. Prior to developing a monitoring plan, consider what <u>needs</u> to be measured so that you can decide whether your criteria for success have been achieved. This provides a minimum requirement for monitoring.

---

§ For more details, see Section C4 – Logical framework approach, in *Managing Marine Protected Areas: A Toolkit for the Western Indian Ocean* (Box 2.5).

Active restoration is only likely to succeed where effective management is in place, e.g. in Marine Protected Areas (MPAs). Some coral reefs recover naturally once effective enforcement of MPA regulations occurs. Others may be too far down "the slippery slope to slime"[1] when declared as protected areas, or may be too small to recover due to recruitment limitations, or have been ill designed and located, or have suffered years of ineffective management and lost their coral cover since being made "paper parks". Even some well-managed MPAs that have suffered mass-bleaching and coral mortality have failed to recover after a decade and remain dominated by macro-algae. In some such cases, active restoration may be appropriate to assist recovery and be undertaken as a part of MPA management.

Luckily, there is much helpful information for MPA managers available free on the internet, which gives excellent advice on planning and monitoring, and can be utilised by those considering rehabilitation. Particularly useful in the context of planning and carrying out a rehabilitation project is the *Managing Marine Protected Areas: A Toolkit for the Western Indian Ocean* (IUCN, 2004)[2]. Most of the management advice applies equally well in all tropical oceans and complements and expands on that given here. Advice comes in the form of succinct two-page briefings ("theme sheets") on each topic with links to further information. The down-to-earth advice is focused on MPA management but much applies to rehabilitation project management.

The Toolkit has also been adapted for South Asia [Download at: www.southasiamcpaportal.org/toolkit/]

Pertinent sections (theme sheets) include:

A2: MPA goals and objectives – general advice on setting goals and objectives.

A5: Integrated coastal management – the role of ICM and the role of MPAs in ICM.

B1: Participatory techniques – ways of actively involving stakeholders.

B2: Conflict resolution – ways of dealing with issues such as conflicting aims among stakeholders.

B4: Local and traditional knowledge – making use of local knowledge in the planning phase.

C3: Management plans – advice on management plan preparation.

C4: Logical framework approach – insights into "logframes", indicators, means of verification, etc.

D2: Consultants and experts – how to deal with these if you need them.

D3: Partnerships and volunteers – how to increase your capacity using partnerships and volunteers.

E1: Financial planning – advice on financial plans and estimating costs.

E2: Financial management – advice on budgeting.

F8: SCUBA & snorkelling equipment – issues of safety and maintenance.

F9: Moorings and buoys – advice on installation to avoid anchor damage.

G1: Monitoring and evaluation principles – uses of M&E in management and reporting.

G3: Monitoring coral reefs – introduction and sources of further information.

G5: Monitoring physical conditions – basic advice and links to further information sources.

G6: Socio-economic monitoring – principles of quantifying benefits of project to communities.

G9: Assessing management success – principles of assessing success against objectives.

H5: Biodiversity & ecosystem health – explanation of these concepts in relation to management.

H6: Coral reef rehabilitation – succinct and sound advice on which we expand in this manual.

H7: Coral bleaching – issues of resistance and resilience, and monitoring and mitigation.

H8: Crown-of-thorns outbreaks – advice on monitoring and how to respond to outbreaks.

This publication is freely available at URL: cmsdata.iucn.org/downloads/mpa_toolkit_wio.pdf and can be viewed on-line at: www.wiomsa.org/mpatoolkit/Home.htm

See www.wiomsa.org/mpatoolkit/Links.htm for links to the references in the theme sheets. These direct you to websites where the references can be downloaded or viewed for no cost. Some links are to html documents but the majority are links to on-line pdf files.

---

1. Pandolfi, J.M., Jackson, J.B.C., Baron, N., Bradbury, R.H., Guzman, H.M., Hughes, T.P., Kappel, C.V., Micheli, F., Ogden, J.C., Possingham, H.P. and Sala, E. (2005) Are U.S. coral reefs on the slippery slope to slime? *Science*, 307 (5716), 1725-1726.

2. IUCN (2004) *Managing Marine Protected Areas: A Toolkit for the Western Indian Ocean*. IUCN Eastern African Regional Programme, Nairobi, Kenya, xii + 172 pp.

In order to measure success (or to catalogue failure), you will need some form of monitoring [Stage 5]. Monitoring requires time and effort and resources and should therefore be focused on just those measurements that are needed to assess the progress towards objectives and success criteria (unless willing marine biologists are available at no cost to do more detailed surveys). As well as tracking the progress of the project towards objectives, regular monitoring allows adaptive management, and the early identification of developing risks such as bleaching, disease and outbreaks of coral predators (e.g. Crown-of-thorns starfish or coral eating snails). "Monitoring" should include both regular systematic surveys to measure progress of the project and routine checks on the coral transplants and conditions at the rehabilitation site. The former might be scheduled at 6 or 12 month intervals, whereas the latter might be carried out at one or two week intervals. The checks can quickly identify potential problems and initiate troubleshooting or adaptive management responses.

For active restoration, measuring success can be made easier if you set up a number of "control" or comparison areas at your degraded site where no active interventions are carried out. The control areas have to be in the same habitat and be exposed to the same environmental conditions as the rehabilitation site if you are to make valid comparisons. You can then compare what happens over time in areas where you have actively assisted natural recovery processes, and what happens in adjacent areas where you have just let natural recovery (if any) take its course. The costs are what you've paid out (see Chapter 7 for a discussion of ways of costing restoration); the benefits are any improvements of indicators (e.g., % live coral cover, numbers of fish grazers, rates of coral recruitment, increases in fish biomass, enhanced biodiversity) in restored areas over and above those in the control areas. Given the increasing amount of reef degradation, the high costs of active restoration, and the potential benefits in terms of

learning lessons from projects that include an element of experimental design, such an approach is strongly recommended wherever possible. The presence of control areas is also a good way of demonstrating to stakeholders that they are achieving success by their efforts. The time-span over which changes are evaluated should be at least several years to match the expected time-course of recovery. Studies show that natural recovery usually takes at least 10 years. Long-term (10 years +) restoration is the goal, not short-term, often ephemeral, improvements in indicators.

For mitigation projects, where it is important to show that the translocation of corals has not caused undue damage, we recommend setting up at least three control sites on adjacent healthy reefs in a similar environment. If a disturbance such as a bleaching event or predator outbreak does equal damage to the control sites, then the mitigation exercise cannot be blamed for the loss of coral.

The three major outputs of the stage 3 planning process should be:

1. A feasible time-bound transplantation plan (and, if needed, a rearing plan).

2. A feasible plan for monitoring the success or otherwise of the rehabilitation project.

3. Measurable and objectively verifiable time-bound criteria for success that have been agreed among stakeholders and which derive from the broad aims of the initial scoping and more detailed objectives set out in the rehabilitation plan.

2.5 Implementation of rehabilitation plan and monitoring [Stages 4 and 5]

Methods of carrying out reef rehabilitation are discussed in Chapters 4–6 but Figure 2.5 provides an overview of the steps and decisions that may need to be taken. There are two main paths: firstly, where wild stock is used to provide transplants (either relatively small numbers needed for a small area, or a mitigation project where there are plenty of "rescued" corals), and secondly, where a nursery stage is needed to prevent unacceptable collateral damage to donor reefs. In rare cases, nurseries with appropriate species from appropriate environments may already be in existence but in general you are likely to need to rear transplant material that is tailored to the target rehabilitation site.

Key points to note are:
1) the use of either first generation transplants (4.1.4) or first generation nursery reared colonies (4.2.6) as source material for additional transplantation if this is required, and
2) the use of monitoring (5.1) to inform adaptive management (e.g. further transplantation, change of species if some survive very poorly) and give feedback to stakeholders. Clearly, it is important to ensure a diversity of source colonies (a recent paper[5] suggests at least 30 genotypes might be a good starting point) or corals of

A *Porites cylindrica* colony with White Syndrome on Luminao reef, Guam (L. Raymundo).

 To reduce risk, transplant a broad cross-section of appropriate species (or growth forms) with due regard to both the "reference site" and availability of source material.

 Use your "reference site" to guide the density at which you transplant, remembering that a doubling of density may quadruple costs.

 To reduce risk, transplant corals in well-separated patches within the rehabilitation site.

 Work with natural recovery processes and think long term.

 At every stage make sure that your plans match the financial and human resources available.

 To minimise impacts on the natural reef you can use some of the nursery reared or "first generation" transplants (once grown) as sources of fragments for further cycles of rearing and transplantation.

 Monitoring should include both regular systematic surveys to measure progress of the project and routine checks on the coral transplants and conditions at the rehabilitation site.

 Focus your systematic monitoring to 1) track progress towards objectives, 2) allow you to evaluate success and 3) provide feedback to stakeholders.

 Carry out frequent routine checks on transplants so that potential problems can be identified early and adaptive management undertaken.

 Set up "control" sites, where no active interventions are carried out, in order to assess whether it is natural recovery or your active rehabilitation that has led to improvements in indicators at the site. Conversely, in mitigation projects, set up control sites on adjacent healthy reef to show whether losses of coral are due to translocation or external factors.

opportunity, particularly if reared coral material is to be used to generate further transplants. However, remember that you are not trying to create a habitat but trying to create the conditions to allow it to recover naturally. Thus surrounding reefs should eventually supply more diversity once natural recovery processes are functioning better.

Staff of the Parque Nacional Arrecifes de Cozumel teaching visiting school children about marine conservation before guiding them around the reef (C. Martinez Ceja).

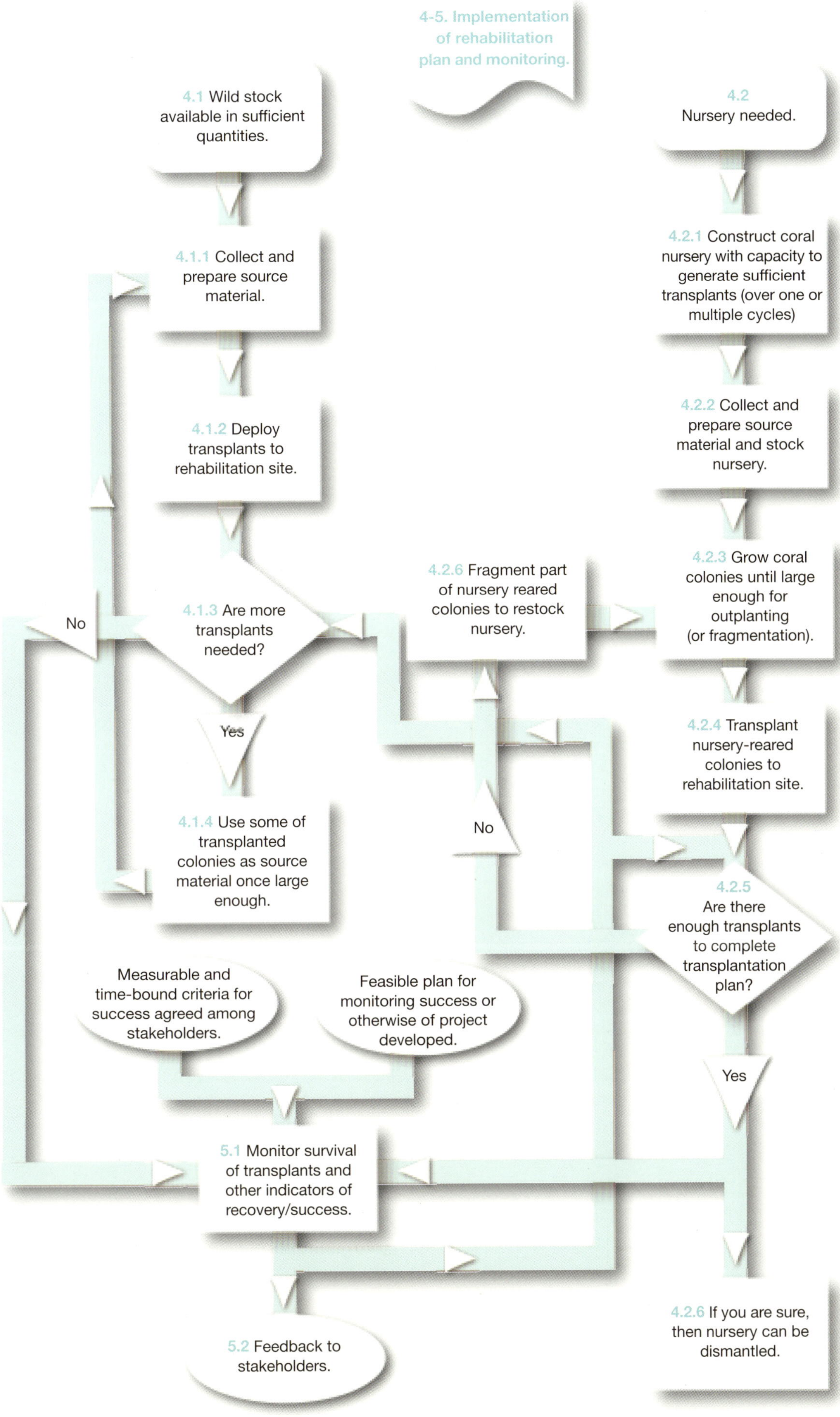

Figure 2.5. Steps and decisions in implementing a reef rehabilitation plan, monitoring and adaptive management.

References

1. Clark, J.R. (1996) *Coastal Zone Management Handbook.* CRC Press Inc., Boca Raton. 694 pp. ISBN 978-1-56670-092-4

2. Golbuu, Y., Victor, S., Penland, L., Idip Jr, D., Emaurois, C., Okaji, K., Yukihora, H., Iwase, A. and van Woesik, R. (2007) Palau's coral reefs show differential habitat recovery following the 1998-bleaching event. *Coral Reefs*, 26, 319-332.

3. Raymundo, L.J., Maypa, A.P., Gomez, E.D. and Cadiz, P. (2007) Can dynamite-blasted reefs recover? A novel, low-tech approach to stimulating natural recovery in fish and coral populations. *Marine Pollution Bulletin*, 54, 1009-1019.

4. Wilkinson, C. (2008) *Status of Coral Reefs of the World: 2008.* Global Coral Reef Monitoring Network and Reef and Rainforest Research Centre, Townsville, Australia. 296 pp. ISSN 1447-6185
[Download at:www.reefbase.org/download/download.aspx?type=1&docid=13311]

5. Shearer, T.L., Porto, I. and Zubillaga, A.L. (2009) Restoration of coral populations in light of genetic diversity estimates. *Coral Reefs*, 28, 727-733.

6. Baums, I.B. (2008) A restoration genetics guide for coral reef conservation. *Molecular Ecology,* 17, 2796-2811.

7. Wabnitz, C., Taylor, M., Green, E. and Razak, T. (2003) *From Ocean to Aquarium.* UNEP-WCMC, Cambridge, UK. 64 pp. ISBN 92-807-2363-4
[Download at: www.unepwcmc.org/resources/publications/UNEP_WCMC_bio_series/17.htm]

8. Hill, J. and Wilkinson, C. (2004) *Methods for Ecological Monitoring of Coral Reefs: A Resource for Managers. Version 1*. Australian Institute of Marine Science, Townsville, Australia. vi +117 pp. ISBN 0 642 322 376.
[Download at: data.aims.gov.au/extpubs/attachment Download?docID=1563]

9. English, S., Wilkinson, C. and Baker, V. (1997) *Survey Manual for Tropical Marine Resources.* 2nd Edition. Australian Institute of Marine Science, Townsville, Australia. 390 pp.

Chapter 3.

Managing risks in reef restoration projects

Overview of risk assessment

Assessing the most relevant risks

Five-step process for prioritising and managing risks

Mitigating risk and adaptive management responses

David Fisk and Alasdair Edwards

3.1 Introduction

What threats and risks can be avoided to improve the chances of success of a restoration effort? What are the temporal, spatial, qualitative and quantitative factors that are most important when assessing risk factors? One approach is to be aware of what has caused significant problems in other projects and to ensure that the same compromising factors are accounted for in the planning and design stages. The limitation of this approach is that there is no specific set of risk factors that will be present for every project, as each restoration site and project have their own unique set of circumstances. For example, in the case of ship groundings on reefs, there can be unique factors that need to be addressed which are characteristic of ship impacts, like the production of large volumes of rubble and the presence of potentially toxic anti-fouling paint[1-3]. It has been suggested that in reef systems where there is evidence of a high rate of natural coral recruitment, allocating limited resources to managing the most obvious sources of disturbance may be a better approach than active rehabilitation, leaving natural regeneration processes to restore the damaged reef[4-5].

Previous rehabilitation projects (including some of the case studies presented in Chapter 8) show that the range of issues that can negatively influence the success of a project are numerous and diverse. Many of the adverse factors that come into play are unexpected[6] or not adequately accounted for in the initial planning stages or in the project design. It is clear that some of these negative influences could have been considered in the planning stages (during scoping and choice of site) if a more rigorous risk assessment had been carried out.

This chapter formalizes the lessons learnt from past studies, by presenting a structured approach and relatively simple assessment protocols so as to minimise risks in future projects. If you undertake effective risk management at the planning stage, and incorporate appropriate responses to the perceived risks in your project design, you should have a better chance of success than many projects in the past.

3.2 Overview of risk assessment

At the outset of the risk assessment process, it is necessary to distinguish between locally manageable risks (local human impacts), and externally derived threats to project success that cannot be managed locally. Nonetheless, threats that are external to a project's immediate influence (e.g., global climate change[7], tropical cyclones, tsunamis) should be considered in a project design so as to mitigate likely impacts. At the very least, such threats should be included explicitly among the assumptions made when setting project objectives.

In addition, it is critical that a cycle of monitoring and adaptive management[8] is incorporated into all projects to help to reduce the risk of failure. That is, most of the aims of a project should treat the project management cycle as a process that occurs over an ecologically meaningful time scale (e.g. 10 or more years) and not as a short term transplantation event (though there are some exceptions to that rule).

Within the project management cycle there needs to be an effective monitoring and maintenance regime to reduce risks such as competition from algae and predation. Appropriately scheduled maintenance and monitoring throughout the project life can provide early warning of problems and trigger adaptive management responses when necessary.

Macroalgae infested Acropora (left) and Pocillopora verrucosa (right) in the Funafuti lagoon (Tuvalu) close to the main township. Indicators like these suggest eutrophication due to nutrient input from the township and/or a lack of herbivores possibly as a result of overfishing (D. Fisk). A survey of other parts of this lagoon indicated that the high macroalgal cover in competition with live coral was restricted to parts of the lagoon adjacent to human habitation, suggesting localised anthropogenic influences on the lagoon ecology. Such information is important in addressing community concerns over the health of their reef. It also suggests that restoration efforts should be focused on the causes of these impacts (passive restoration) rather than undertaking active restoration in these areas.

What is the role of risk management in reef rehabilitation?

Risk management is predominantly the practice of systematically selecting cost-effective approaches for minimising the effect of environmental disturbances and threats (e.g. predation) to a restoration effort. Be aware that all risks can never be fully avoided or mitigated (Figure 3.1). As a consequence, all projects will have to accept some level of risk.

The risk of disturbance or threat refers to a combination of the probability or frequency of occurrence of a disturbance and the magnitude of its consequences.

To further complicate the risk assessment process, you need to be aware of, and perhaps take into consideration in your assessment of risk, factors that may not be immediately apparent with respect to a specific disturbance. These risk factors can be reasonably predicted to occur as a consequence of the initial disturbance, but may only become an issue some time later. For example, following a Crown-of-thorns starfish (*Acanthaster planci*) outbreak, a few large individuals often remain in the area but may not be in the impacted site where you want to apply restoration efforts. After introducing new colonies to the denuded site, the starfish can be attracted to this new source of food and seriously affect your transplants.

Crown-of-thorns starfish (Acanthaster planci) eating a Pocillopora colony in Hawaii (K. Kilfoyle). Interestingly, at a site near Bolinao in the Philippines more transplanted corals were lost to Crown-of-thorns (COT) predation during a warming event in 2007 than to bleaching. The site had never had a COT outbreak in living memory, emphasising that you should always expect the unexpected when attempting reef rehabilitation.

Message Board

- Risk management is a structured approach to manage uncertainty related to (i) potential threats and disturbances to rehabilitation projects and (ii) the lack of scientific knowledge about reef restoration.

- The potential impact of uncertainty can be mitigated by embedding both monitoring and adaptive management in the project design.

- Managing risk involves using past experience from other studies, applying those lessons to your project, and being aware that the unexpected is always possible.

- Potential risks and uncertainties should be explicitly communicated to stakeholders and funders at the planning stage.

- Each project will have a unique set of environmental and socio-economic conditions, such that known risks will vary in their potential impact on the outcome of a project.

- At the very least, you can use the lessons learnt from past projects in a structured way to assess the likelihood of the occurrence of known risks, as well as their expected importance to your project.

- Clearly defining the aims of your restoration project and understanding the temporal and spatial implications of those aims is a critical initial step in managing the potential risks to your project.

- A proper monitoring plan is central to both adaptive management and risk mitigation in a rehabilitation project (not an optional extra).

- A key lesson from past *active* restoration projects is that you should expect the unexpected.

Figure 3.1. The main classes of risk and general management responses.

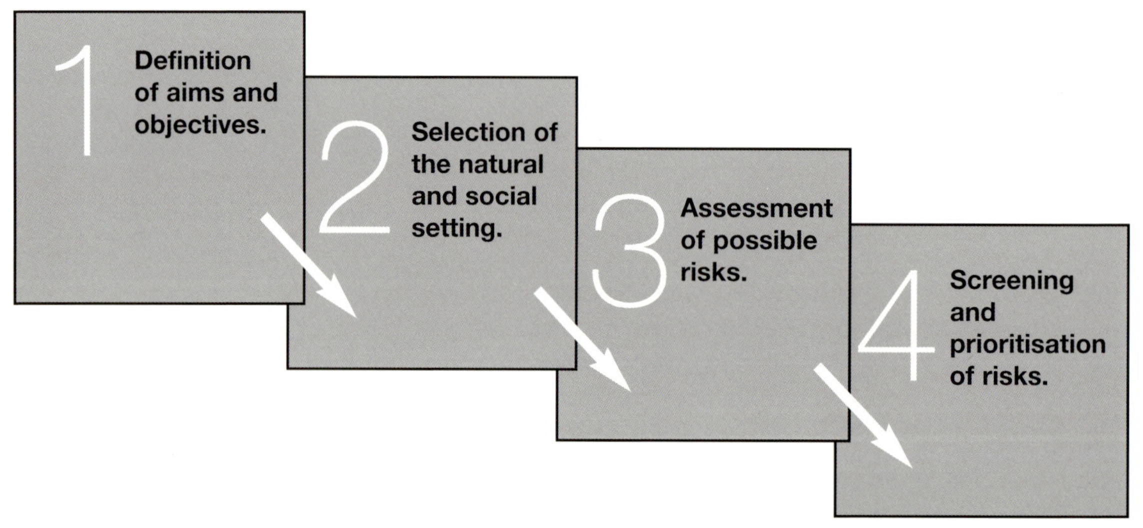

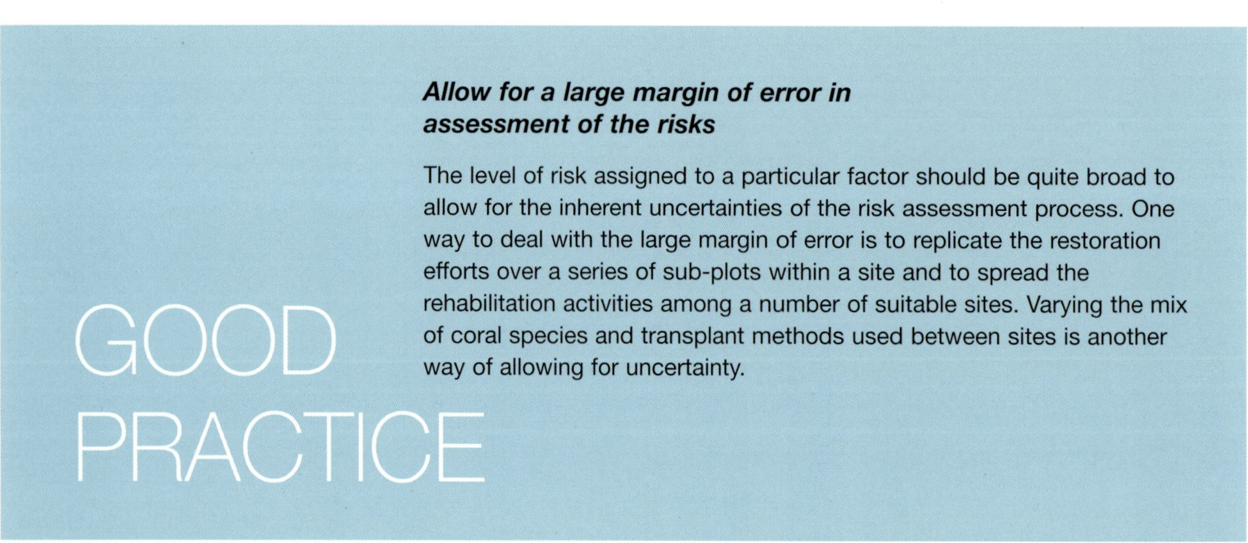

Allow for a large margin of error in assessment of the risks

The level of risk assigned to a particular factor should be quite broad to allow for the inherent uncertainties of the risk assessment process. One way to deal with the large margin of error is to replicate the restoration efforts over a series of sub-plots within a site and to spread the rehabilitation activities among a number of suitable sites. Varying the mix of coral species and transplant methods used between sites is another way of allowing for uncertainty.

GOOD PRACTICE

Figure 3.2. Hierarchy of risk assessment to incorporate into your project design.

3.3 Assessing the most relevant risks during project design

A risk management strategy should incorporate four essential hierarchical components (Figure 3.2).

1. Definition of aims and objectives

Carefully define the aims and objectives of your rehabilitation project (if a location is already assigned). Alternatively, if the aims are already assigned, then carefully choose the location so that they may be achievable. The original aims of a restoration project will determine the range of risks that you need to take into account. Many of those potential risks can affect a number of different project aims. In general, restoration primarily endeavours to improve ecosystem structure and function in degraded reef areas. Your project objectives may range from *passive* indirect measures, to *active* direct measures, all with the purpose of redressing certain defined aspects of reef degradation or damage.

2. Selection of natural and social setting

Once a location and project aims are established you need to consider carefully the natural and social setting of the proposed rehabilitation site. This will refine the range of risks that you need to consider in a rehabilitation plan. In most instances, you will need risk predictions relating to the ecology of the site, and to the natural and anthropogenic threats, which will be unique for a particular site. Management of risks will depend also on the social context of the restoration site and your project aims. "Social context" refers to the human influences and decision-making structures in place at the location, specifically, the degree to which the key stakeholders are involved, and are in agreement on, the restoration aims and activities.

3. Assessment of possible risks

A full risk assessment needs to be an integral part of the project design phase and should be carried out early on so that the outcomes can be built into the implementation phase (Chapter 2). Scoping and field assessment by a specialist is strongly recommended during the design phase. Risk factors should be collated and assessed at the initial scoping stage by the team undertaking the project (a team that ideally includes a reef ecologist, project managers, decision-makers, and local community members). Critical environmental and biological factors that may impact your project need to be assessed via the input of an experienced reef ecologist who is able to interpret the environmental signs that are present at the site selected for rehabilitation. This will include an assessment of previous environmental and human usage trends of your site. Assessing and interpreting trends in the past history of a site should help guide the planning of the rehabilitation project and allow you to make it more robust to disturbances.

4. Screening and prioritisation of risks

The management of risk requires that you compile as much information as possible on biological, environmental, and social risk factors that relate to your site and prioritise these to arrive at a strategy of where, when, and how to proceed (which can also include the decision not to proceed). The iterative nature of the risk management process may mean that factors originally classed as low risk could become high risk factors during the project's life. We strongly recommend that the risk management plan includes provision for regular reviews of progress and conditions at the site. This allows the project to change emphasis and methodology if necessary in order to achieve its aims, i.e. your project design needs to be flexible and reactive to changing threats[8].

Ideally, you should follow a prioritisation process whereby the risks with the greatest loss and the greatest probability of occurring are addressed first, and risks with lower probability of occurrence and lower resultant loss are dealt with in descending order. In practice, the process of prioritisation can be very difficult to complete, and balancing between risks with (1) a high probability of occurrence but lower loss, and (2) a high loss but lower probability of occurrence, can be difficult.

3.4. Five-step process for prioritising and managing risks

To manage risk, a structured response related to perceived threats can minimise the chance of failure of a restoration project. The response strategies include:

- avoiding the risk,

- mitigation strategies for reducing the negative effect of the risk, or reducing the magnitude of loss, or probability of occurrence,

- spreading the risk among other components of the project (by replication of effort in different spatial, species composition, and temporal settings, e.g., by spreading

A lone Acropora *coral colony on a macroalgae dominated reef (N. Graham). Unless the broader management issues which have allowed the seaweed to dominate are resolved first, there is high risk that active restoration measures will fail.*

the risk among different habitats, or between different species groups or source areas, or over different times of the year or different environmental conditions), and

- accepting some or all of the consequences of a particular risk and budgeting for that factor.

An initial planning and assessment exercise where you predict the risks to the project and plan how you will manage these risks, should be based on local, regional, and global knowledge, and should include the following five steps.

1. Setting out your rehabilitation project's specific aim(s). Once you have established and agreed your aim(s) with all stakeholders, then specific spatial and temporal requirements for the project will result from the aim(s). The assessment of risk involves a proper understanding of the spatial and temporal scale implications of each aim.

2. Defining the risks associated with the natural (biophysical) and social setting of the proposed rehabilitation site.

3. Assessing and prioritising the risks to your project early on during the design phase (listing, assigning a perceived probability of occurrence and magnitude of impact (i.e. consequences) to each risk, and then prioritising).

4. Development of response options for mitigation of risks using realistically available technological, human and organisational resources.

5. Integration of earlier steps to develop a Risk Management Plan.

Step 1. Setting out your project's aim(s), and understanding the spatial and temporal implications of those aims

The conditions under which your project will operate are partly determined by your aims and objectives, and partly by the social and ecological circumstances of the site chosen for restoration. Often a project will have several aims and objectives but for simplicity, each major aim will be treated separately here. You will need to assess and prioritise any conflicting risk factors when more than one main aim is identified for a single project.

Risk management considerations relating to the most common aims (see Box 3.1) are outlined below.

Box 3.1 Rationales and aims of reef rehabilitation.

Common reasons for carrying out reef restoration interventions include:

- Lack of awareness in a local community and a poor appreciation of the economic and cultural value of reef ecosystems.
- Loss of biodiversity.
- Loss of productivity (food species).
- Loss of key reef components (usually coral, but also adjacent seagrass or mangroves) due to natural disturbances (bleaching, storm damage, coral predation and disease),
- Loss of key ecosystem processes (e.g. recruitment of juvenile corals, grazing of macroalgae by herbivorous fish or urchins) and services.
- Provision of alternative livelihoods (e.g. culture of aquarium products, tourism) for stakeholders who agree to stop harvesting reef resources.
- Mitigation for developments that will adversely impact coral reef species at a site, especially the relocation of threatened corals.

Common aims that previous restoration projects have cited as their motivation include:

1. Building public awareness and environmental education.
2. Promoting recovery of biodiversity.
3. Increasing biomass and productivity.
4. Assisting recovery of key reef species or ecosystem processes.
5. Development of alternative livelihoods.
6. Mitigation of damage or degradation.

1. Building public awareness and environmental education

Often the aims of raising awareness of coral reefs and education of local communities result in demonstration projects that have to be located at readily accessible sites situated in areas where communities can claim ownership of the activity and/or where some degree of surveillance of the site can be done by that community. The objective is to have high community visibility and awareness of the project.

Risk management messages

- There is a risk that practical considerations may take precedence over ecological considerations, leading to poor restoration outcomes due to low attention to ecological requirements.

- Although stimulating community involvement in a project is very important, you need to be careful that this aim does not jeopardize other aims, that is, multiple aims can be hard to manage effectively as the pursuit of one aim can negatively influence the success of another (e.g. Ch 8: Case study 2).

- Awareness/education projects are usually small in spatial scale and do not generally last for more than a couple of seasons or years. While building awareness, they therefore tend to have poor outcomes in terms of real reef rehabilitation. As a consequence, we would advise that this aim should generally be adopted without linking other (e.g. ecological) aims with it. There is, of course, a danger that poor ecological outcomes will be counterproductive in terms of public engagement, so the consequences of ecological compromises driven by "public awareness" convenience need to be considered very carefully.

- On the other hand, large long-term rehabilitation projects not necessarily constrained by the practical requirements of education and awareness building, can deliver education and awareness benefits at appropriately accessible sites (e.g. Ch. 8: Case study 9).

2. Promoting recovery of biodiversity

A biodiversity protection aim using *active* restoration, requires that your project design will result in a functional and diverse reef community. Often, an implied outcome is that reef resilience will be enhanced through fisheries management or through the preservation of a specific coral species. Conversely, *passive* restoration to control fishing and build herbivore populations[9] can markedly increase resilience.

Risk management messages

- Sites that are most ecologically suitable for enhancing and promoting biodiversity may be difficult to work and can entail high logistical costs.

- Biodiversity preservation is a long-term aim that requires long-term management and maintenance to be successful.

- *Active* restoration efforts cannot automatically build a resilient community though in theory, they can enhance resilience through the establishment of a diverse range of healthy mature colonies. It may take decades before a highly diverse and resilient community can be established.

Drupella cornus snails feeding on a plate of *Acropora hyacinthus* in the lagoon at Funafuti (Tuvalu). Inset: *Close-up of a group of Drupella and their feeding scar on an Acropora florida colony (D. Fisk). Careful examination of feeding scars and adjacent areas is necessary to distinguish between Drupella, disease, fish bites, Crown-of-thorns starfish and other causes of lesions. Note the rapid reduction in live coral tissue (evidenced by the fresh white scar area with no sign of colonisation by turf algae). High incidences of coral predators are a significant risk to restoration efforts and can be difficult to manage, generally requiring constant vigilance and maintenance efforts.*

3. Increasing biomass and productivity

A productivity aim usually relates to a focus on fisheries enhancement by restoring or creating suitable reef habitat for commercial or subsistence fishery species (both vertebrate and invertebrate). Some previous projects have focused on increasing habitat complexity and providing refuges for fishery species by active restoration (e.g. Ch. 8: Case study 2) whereas enforcement of no-take areas (passive restoration) can also generate significant rises in fish biomass without habitat modification (e.g. Ch. 8: Case study 10).

Risk management messages

- In general, meaningful gains in biomass or productivity are rarely achieved by active restoration techniques because of their small spatial scale. Such techniques cannot be applied at a suitably large spatial scale, i.e. at scales orders of magnitude larger than the current practical limit of restoration techniques (1–10 ha).

- You will need to incorporate a medium to long-term project timeline (3+ years, depending on the species involved) to achieve any tangible measure of success, as multiple recruitment seasons are usually necessary to allow for build up of fishery biomass. The implications are that effective management input over the long term will be required also.

- In contrast, passive restoration techniques (effectively managing the drivers of reef degradation) can be reasonably expected to result in productivity gains given sufficient time. However, depending on the main sources of degradation that were present, any productivity gain may be highly variable among potential sites[10].

4. Assisting recovery of key reef species or ecosystem processes

This aim is usually achieved by *passive* restoration efforts that are expected to result in increased natural recruitment or survival rates of corals, fish or other reef species. The assumption is that natural build up of key reef species will occur under favourable environmental conditions (e.g. the absence of disturbance factors), and in time will result in higher recruitment and survival rates, and eventually a more biodiverse and resilient reef system. It may be assisted by *active* restoration (e.g. transplantation of *Acropora palmata* in Case studies 6 and 9 in Chapter 8).

Risk management messages

- Restoration of key reef species may initially be at a small spatial scale (e.g. by focusing on managing key habitat for selected species) with the expectation that those small areas will then enhance recruitment elsewhere in a reef system. This is difficult to achieve. For example, key habitat may be fish spawning aggregation sites or naturally high coral diversity areas that are expected to supply recruits to other areas. The rationale for this approach assumes there will be adequate connectivity and larval dispersal, that larval survival will be adequate, and that other natural processes will be favourable to recovery.

- The project design and resourcing need to reflect the likely timescale of recovery; this may vary from 5 years to decades depending on the severity and spatial scale of the impacts.

5. Development of alternative livelihoods

This aim usually entails an agreement with local communities or resource owners for restricted harvest regimes in exchange for income from (i) fees to resource owners, (ii) employment in tourism (e.g. as guides, boatmen), (iii) aquaculture, or (iv) coral farming to provide products for the marine aquarium trade or support re-establishment of selected species into natural populations or tourism related habitat enhancement. The aquaculture of highly prized giant clams (*Tridacna* spp.) for food or as an aquarium species is an example.

Restoration techniques can also be employed to accelerate recovery at or enhance sites that are already tourist attractions, are being developed for tourism or have been damaged during resort construction (Ch. 8: Case study 3). Examples are sites, which may have been damaged by storms, predation, disease or bleaching events, that are used as snorkel trails or underwater viewing areas by glass-bottomed boats. Sites that are adjacent to a tourist resort also can provide added economic value to the resort if attractive reef habitat is made readily accessible to guests.

Risk management messages

- You may need to consider establishing aquaculture activities at sites, which although ecologically suitable for the chosen species, may be logistically challenging.

- The growth rates of species (corals, clams, urchins, etc.) being cultured for the aquarium trade or for re-establishing depleted natural populations, will determine the minimum time scale needed. You need to ensure that the time-scale of budgetary support matches the aquaculture cycle.

- If proposed aquaculture involves use of feeds for fish, then beware of potential eutrophic impacts.

- When multiple aims are adopted, conflicting species requirements may occur. For example, the demands of aquarists (who require small coral colonies of very specific species, colours, etc.) will be different to those of managers aiming to enhance natural populations. Thus, two complementary sets of nurseries, one for income generation, one for rehabilitation might be needed. These issues need to be explored at the project design stage.

- Enhancing or accelerating recovery at a site with high tourism value means that the choice of site is predetermined (rather than being selected as part of the project design phase) with the risk that the site may not be optimal in terms of likely restoration success. Critical parameters to consider include: allowing sufficient time for attachment and growth of transplanted corals (1–2 years), and planning for longer periods (3–5 years) before gradual natural accumulation of fish and other organisms from natural recruitment processes are noticeable in the restored area (in contrast to short-term build up of fish abundance via the attraction of fish already in the vicinity, e.g. through the introduction of artificial reef structures).

6. Mitigation of damage or degradation

As stated in Chapter 1, mitigation of damage refers to the reduction or control of the adverse environmental effects of a project, but it also includes restitution for any damage to the environment through replacement, restoration, or creation of habitat in one area to compensate for loss in another. This often involves moving corals and other organisms from a designated high disturbance site (usually due to a development) to an adjacent site outside the development impact zone. Relatively short time scales (2+ years) are probably required to assess the survival (success/failure) of the transplantation. Success will depend on how well the transplants adapt to the new site, and where necessary, whether there has been adequate attachment.

Risk management messages

- When re-locating corals and other organisms you should try to re-create the spatial arrangement (paying particular attention to zonation and depth) and density of organisms that existed at the source site and ensure that the receiving environment is compatible (e.g. with respect to current/wave exposure) with the original one. To do this you need to find a relatively bare or previously degraded site that has biophysical conditions as similar as possible to the original source site. However, finding suitable sites for relocating corals can be difficult as logically, coral communities would be expected to already exist at similar non-impacted sites. Thus, by implication, relatively bare areas may well be unsuitable for reasons that are not clear (e.g. susceptibility to decadal or longer-term disturbances).

Step 2. Defining risks associated with the natural and social setting of the rehabilitation project

To define the risks associated with the *natural setting* you need a full description and analysis of the rehabilitation site, including key factors that influence the physical and biological processes that shape the reefscape. The major sources of natural or environmental risk and the main risk factors associated with each, along with predictive threats and management responses, are outlined in Table 3.1. Those related to the natural setting include:

1. History of natural and anthropogenic disturbances at a site,

2. Connectivity and spatial relationships of a site with respect to the hydrodynamic regime (tidal characteristics, marine and coastal ecosystem links, terrestrial links, regional marine connectivity), and

3. Coral transplantation issues (sources of coral transplants and potential collateral damage, transplant species and growth forms, life history and reproduction).

To define the risks associated with the *social setting* you need to identify all human related factors concerned with governance, decision making, and ownership issues affecting the site. The major sources of social or human associated impacts and the main risk factors associated with each, along with predicted threats and possible management responses, are outlined in Table 3.1. These can be split into:

4. Social and political setting – local social and political factors (site selection implications, local decision making and management arrangements, stakeholder understanding, unpredictable factors, effects of local economic changes, protection, post-funding stakeholder issues), and

5. Management issues – administrative considerations (adaptive measures, training and capacity of personnel, stakeholder engagement, monitoring and reporting protocols).

Coral nursery platform being assembled during a community restoration project in north-western Luzon, Philippines (R. Dizon).

Table 3.1 This table presents an outline of how you can define and respond to risks to your reef rehabilitation project. It lists sources of risk and details of major factors associated with each source, the kind of information you should seek about potential threats, risk predictions associated with various factors, and finally management responses that can be implemented to mitigate the most likely risks. To make the most use of this table you should systematically address each of the main sources of risk, along with the listed factors to consider with each source. For each factor, consider the various risk predictions in terms of their likelihood at your site and damage they might do to your project, then look for guidance and suggestions in the "Mitigation or adaptive management response" column to ensure you are maximizing the likelihood of success. You should repeat this exercise for each project because different locations and even sites within a location may have different circumstances and risks to consider.

General information needs: As part of the design phase of your rehabilitation plan (Ch. 2: Figure 2.1), you should gather information on potential risks to project success at your proposed site(s). The aim of the information gathering exercise is to estimate the likelihood that your planned rehabilitation will be compromised by the various groups of factors below during the lifetime of your project. The level of risk will thus be dependent both on the objectives and timeframe of the project (see Chapter 2). If risks appear significant, then you need to decide on mitigation or adaptive management responses (right-hand column) to minimize them. If mitigation has a poor chance of success then you might wish to consider changing objectives (or site) in discussion with stakeholders, or consider abandoning the project if no viable alternative site can be found.

1. Site history of disturbance (natural and anthropogenic)

1.1 Physical factors: tropical cyclones and storms, freshwater runoff and groundwater seepage, warm water anomalies that cause coral bleaching and mortality.

    Information needs: You should collate information on past physical disturbances and impacts at your site (or nearby reefs, if nothing known) and look for evidence of trends (e.g. increased frequency due to climate change) or cycles (e.g. seasonal fluctuations, or El Niño Southern Oscillation [ENSO] events occurring every few years). Make full use of local knowledge such as that of fishers as well as published data such as weather records. Focus on finding out what physical and chemical factors are mostly likely to stress your rehabilitation site and when these stresses are likely to occur.

Risk prediction	Mitigation or adaptive management response
Tropical cyclones and storms: tend to be more frequent at certain times of year (hurricane, typhoon, cyclone or monsoon seasons – see www.aoml.noaa.gov/hrd/tcfaq/G1.html). Newly transplanted corals are more susceptible to detachment than well-established colonies that have had time to self-attach to the substrate. Climate change may lead to more frequent and severe tropical cyclones.	Try to avoid transplanting during the tropical cyclone season (e.g. in western Atlantic: 1 June – 30 November; peak period early to mid-September) or during monsoon season when your site is most exposed to wave action. Where feasible allow a few months for transplants to self-attach[11] before seasonal rough weather. Avoid very exposed sites. Make sure transplants are securely attached to the substratum.
Freshwater run-off and groundwater seepage: may reduce salinity and/or introduce nutrients/sediment/garbage during wettest months of year or during storms (see above). Coral transplants are likely to be stressed and may die in areas subject to such impacts.	Try to avoid sites subject to damaging seasonal (or longer cycle) inputs from land. Where there is freshwater seepage, beware of transplanting to lagoonal sites that appear suitable for corals but are devoid of them (unless cause for absence is evident and not a current threat). Some sites may suffer seasonal sedimentation from run-off or algal overgrowth from nutrient fluxes and have corals adapted to this. At such sites only use locally adapted species as transplants and try to avoid transplanting at times of year when water quality is poor due to run-off from land.

Warm water anomalies: (unusually warm sea temperatures often associated with ENSO events) stress corals and can cause coral bleaching (a symptom of stress) and death if they occur during warmest months of the year and are prolonged (several weeks). Stressed corals appear to be more prone to disease[12] and more susceptible to attack by snail, sponge and other predators. Even in normal years corals may be stressed during the warm season. Warm water stress is predicted to become more frequent and severe as a result of climate change.

Try to avoid transplanting corals during warmest months of the year at your site (e.g. May–September in northern Philippines, March–June in Maldives) and if feasible allow transplants to become established for a few months before sea temperatures reach their annual peak. [*Note:* It may be difficult to avoid both rough weather and warmest months.]

For longer term cyclic events, consult www.osdpd.noaa.gov/ml/ocean/sst/anomaly.html and related websites to check sea surface temperatures (SST) anomalies, HotSpots and Degree Heating Weeks data for your area. These are updated approximately twice a week. If SSTs seem to be unusually warm for the time of year (they may be anomalously high for several months before annual maxima are reached and bleaching occurs), seek advice and consider rescheduling project. Check for ENSO predictions on: www.cpc.ncep.noaa.gov/products/analysis_monitoring/enso_advisory/index.shtml. If there are predictions of an imminent ENSO related warming in your area, consider rescheduling any planned transplantation until after it (this may allow you to identify bleaching resistant genotypes for nursery propagation).

Choose the more temperature tolerant species for your transplants (this may require expert opinion, but if previous bleaching events have occurred in the area, those known to have survived are obvious choices; in general, massive and slower growing forms tend to survive better than fast growing branching, tabulate and encrusting forms). At the very least, use a mix of species that are well-adapted to your site to increase the chance that some will survive a warming event.

If feasible, choose sites with features that help reduce the impact of warm water events (e.g. local upwelling of cooler oceanic water, moderate turbidity (reduces sunlight), good tidal water exchange).

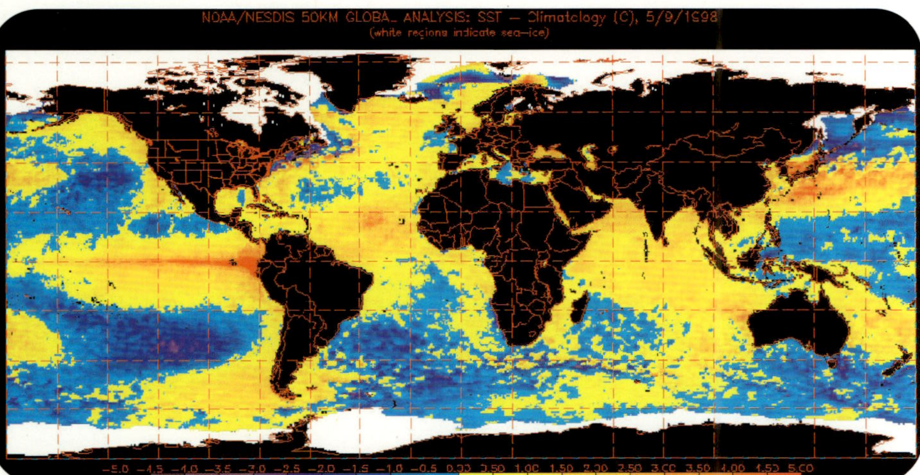

A sea surface temperature (SST) anomaly map from May 1998 from the US National Oceanographic and Atmospheric Administration website at: www.osdpd.noaa.gov/ml/ocean/sst/ (courtesy of NOAA).

1.2 Biological factors: Crown-of-thorns (COT) starfish (*Acanthaster planci*), cushion stars (*Culcita*), gastropod predation (*Drupella, Coralliophila*, etc.), disease, macroalgal dominance and overgrowth, lack of herbivores due to overfishing, sponge and tunicate infestations, fish that attack coral transplants.

Information needs: You should collate information on past (e.g. COT outbreaks – see www.aims.gov.au/pages/reflib/cot-starfish/pages/cot-000.html) and present potential biological threats at your site (or nearby reference reefs) and look for evidence of trends (e.g. increasing accounts of gastropod predation) or cycles (e.g. seasonal fluctuations in disease prevalence, tunicate infestations, macroalgal blooms). Make full use of local knowledge such as that of fishers as well as any published data from the area. Focus on finding out what are the most likely biological threats at your proposed rehabilitation site and whether these are seasonal.

Risk prediction	Mitigation or adaptive management response

Predation: Corals are preyed upon by a range of organisms (e.g. COT starfish, cushion stars (*Culcita*), gastropods such as *Drupella, Coralliophila* and *Phestilla*, and some fish species) and if these predators are in significant numbers, they can seriously damage coral transplants, particularly those recently transplanted and stressed. The main problem is that restoration areas are usually small and easily overwhelmed by predators. These may appear to be present at low density but can be attracted from neighbouring areas by the presence of coral transplants. If there is a recent past history of predator

You should carefully assess the proximity and density of potential predators at the site selection stage. If they are present in significant numbers, there is likely to be a high risk of project failure due to predation of transplants. Mitigation options include daily to weekly or monthly visual monitoring and removal (as needed) of invertebrate predators such as COT starfish and gastropods, but this will involve considerable time and effort (and thus cost). For fish, temporary exclusion of larger parrotfish and wrasse for a week or two by caging of newly transplanted area may assist survival but may be logistically difficult impacts (e.g. COT outbreaks), then future predation (though predators do not appear to be immediately present at the site) may be expected to occur. In MPAs or areas with low fishing pressure, larger parrotfish and certain species of wrasse may be attracted by newly transplanted corals and may both graze and detach transplants.

and, if dislodged, caging may cause damage in a storm. Transplants have recovered from initial intense grazing and appear to become less prone to attack with time. Also, species with denser skeletons appear less prone to fish predation.

Diseases: Corals are susceptible to disease when they are stressed. Warm sea temperatures during the summer months or during warming anomalies (1.1) are correlated with increased coral disease[12-13], thus stressing corals by transplanting them during the warmest months of the year is likely to increase their susceptibility to disease. High nutrients and sedimentation may also be conducive to the prevalence of diseases[12,14].

Black-band disease on a large Diploria *colony at Mona Island, Puerto Rico (K. Kilfoyle).*

Careful handling of corals helps to minimize stress and tissue damage during nursery rearing or transplantation. Careful maintenance of your coral nursery (section 4.5) or transplants to remove any harmful algae or predators should also reduce stress and the chance of infection. Locating your nursery or restoration site where water quality is good should reduce the potential for coral diseases. However, little is known of the source and transmission of coral diseases[15] and once disease is established at a site "treatment" is generally not considered feasible[14]. The *Coral Disease Handbook. Guidelines for Assessment, Monitoring & Management*[14] can assist you in identifying diseases and distinguishing them from tissue loss due to predation, bleaching, invertebrate galls and other non-disease lesions. You should avoid transplanting corals near reefs where diseased corals are prevalent and if disease breaks out in your nursery or transplant site then diseased corals should be removed immediately to reduce the chance of the infection spreading.

Macroalgal dominance and overgrowth: Macroalgae compete with corals for both space and light. They can overgrow, abrade and shade corals and prevent recruitment of corals by pre-empting space[16]. Thus high prevalence of macroalgae is likely to be antagonistic to transplant survival and reef recovery processes.

High macroalgal cover may result from high nutrient loads in the water (due to sewage or aquaculture wastewater, or farm run-off), or from insufficient grazing activity by fish and sea urchins because of overfishing, or from a combination of factors[17]. After natural disturbances that remove corals, algal cover may also rise (but usually only temporarily on healthy reefs). For mitigation, the cause of the high macroalgal cover needs to be determined and appropriate management actions taken to reduce nutrient inputs or reduce fishing pressure at the site. Once these factors are under control, then active restoration might have some chance of success. Management actions may take a considerable time to be effective, so the rehabilitation site may require an initial removal of macroalgae or maintenance to prevent the build-up of harmful macroalgae until there are sufficient numbers of grazers to control algal growth.

Lack of herbivores: Reef herbivores, such as parrotfish, surgeonfish, rabbitfish and sea urchins, serve a vital function in creating space on the reef where invertebrate larvae, including coral larvae, can settle and survive. If there is insufficient grazing then coral recruitment is likely to be reduced leading eventually to reduced coral cover and structural complexity and reef health decline[18].

Reducing the fishing pressure on grazers may allow numbers to recover and grazing to return to a level that allows recruitment and keeps macroalgae in check. But there may be cases where it could be necessary to re-introduce grazing species (e.g. urchins) to areas where they have been very seriously depleted. However, great caution is necessary in any such manipulation.

Sponge and tunicate infestations: Sponge and tunicate (sea-squirt) infestations can lead to significant coral mortality. Some tunicates may occur seasonally and not do much damage but sponges such as *Cliona* (Ch. 8: Case study 6) may overgrow and kill corals, particularly when corals are stressed.

A sponge enveloping a Montastraea *transplant near Cancun in Mexico (K. Kilfoyle).*

Reef substrate immediately adjacent to where corals are to be attached can be cleaned with a wire brush or scraped to remove potentially hostile sessile invertebrates and macroalgae prior to transplantation of corals. Water rich in particulate organic matter such as that near aquaculture farms or where there is run-off from land can favour sponges and tunicates, so management of water quality can reduce the risk of sponge and tunicate infestations. In the short term, monitoring and maintenance of your restored area (or coral nursery) to remove tunicates and sponges that threaten to overgrow corals can help, but this is not sustainable in the longer term without measures to address the root causes of the infestations (e.g. management of water quality at the site).

Movement of corals from donor sites to nursery or transplant sites may lead to the transfer of threats such as disease, predators and invasive algae or sessile invertebrates.	Try to select only healthy colonies or fragments for nursery rearing or transplantation, and remove potentially harmful organisms or damaged parts prior to transplantation.

2. Connectivity and hydrodynamic regime around rehabilitation site

Factors: Water quality, larval transport towards and away from site, hydrodynamic regime.

Information needs: You need to find out as much as you can about the water flows around your site, how these change on a daily and seasonal basis, and how it is connected to surrounding marine areas and influenced by rivers and other land run-off (including anthropogenic discharges). An understanding of the currents, tides, and waves impinging on your site and potential sources of pollution can help you to determine if any of these features will influence the chances of successful restoration. Admiralty Pilots for your area will have much useful data which can be combined with local fisher knowledge and your own observations (e.g. tracking of current flows using low-cost surface drifters[19]). This information together with a knowledge of the life-histories of the corals (e.g. larval competency times – section 5.8) can give you some indication of likely coral larval transport.

Risk prediction	Mitigation or adaptive management response
Water quality: Currents can carry nutrients, chemical pollutants, sediment, low salinity water, and other water quality related disturbance threats to the rehabilitation site.	Ensure that present and predicted upstream activities that negatively impact water quality will not significantly affect your site (with due allowance for seasonal changes in prevailing current flows). Try to select a location for your nursery or transplant site that is unlikely to be impacted by deleterious terrestrial or water borne inputs (either ensure it is distant from potential pollution or upcurrent from it). Avoid sites near sewage outfalls, aquaculture farms, river discharges, etc.
Larval transport: Local currents will determine whether coral and other larvae arrive at your site in sufficient numbers to assist recovery or whether larvae produced by mature transplanted corals at your site will settle locally or be transported away to neighbouring reef areas or to the open ocean. Depending on your project objectives, there are risks that local current patterns may hinder your rehabilitation plans. Isolated and unconnected sites are likely to be less resilient to disturbance.	Try to utilise sites that are connected by current flows to neighbouring reefs and can function as part of a network of reefs that exchange larvae (this is similar to the idea of establishing resilient networks of MPA's[20]). If you intend your site to be a source of future coral larvae for a larger degraded area, then make sure that it is upcurrent of the target area.
Local water flows and tides: The direction and strength of water flows (including water mixing via tides and wave action) and tidal range will affect the chances of successful transplant attachment, growth and survival. Too much water movement may hinder attachment; too little mixing during the warmest months may lead to heat stress and mortality; unusually low seasonal spring tides may expose corals to the air if transplanted too shallow. The hydrodynamic regime at a site may vary dramatically between monsoons or seasons.	When considering whether your site is suitably located take into account seasonal changes. You may have to consider trade-offs between protection from waves during stormy periods and exposure to warming during calm periods in the warmest months.

3. Coral transplantation issues

Factors: Sources of coral transplants, coral species and growth forms, life history and reproduction.

Information needs: A sound knowledge of the biology and ecology of coral reefs is likely to improve the success of your project. In the absence of an experienced coral reef ecologist, there is a large number of helpful documents available over the internet (see lists of references at end of each chapter) and considerable scope for informed judgment using careful planning and a comparable reference site (Chapter 2) to guide your project design. You need to carry out prior detailed inspection of prospective donor sites for corals, proposed nursery or rehabilitation sites, and where necessary a "reference" site (part of the fact-finding for your rehabilitation plan – section 2.3). A reference site is one with a similar environmental setting to the rehabilitation site but which has enough of a coral community left to tell you what ought to survive at the rehabilitation site (if the latter is denuded and the previous coral community is unknown). Work out distances between sites and travelling times and other logistics needed to transport corals.

Risk prediction	Mitigation or adaptive management response
Sourcing coral transplant material: Your site (or sites) chosen for donor colonies (or "corals of opportunity") does not contain sufficient coral for the project without significantly damaging the donor locations.	Reduce the scope of your project and corresponding demand for transplants if alternative donor sites cannot be found with similar conditions to the rehabilitation site. You may also consider using coral nurseries for propagating colonies or fragmenting the first "crop" or transplants once they have grown in order to generate more transplants.
If the distance between your coral source site and nursery or transplantation site is too far to allow relatively stress-free transport (see Chapters 4, 6 and 8) of coral fragments or colonies <u>given your resources</u>, then survival will be poor.	During the planning stage make sure that you have adequate resources to transport required coral source material safely from source site(s) to nursery/transplantation site(s) or consider moving or abandoning your planned transplant site location.
If you use colonies from other habitats (even adjacent but different habitats) you will increase the risk of unacceptable mortality.	Refrain from the temptation to use donor colonies from habitats that are different to the rehabilitation site in terms of wave exposure, depth, water clarity, etc..
Coral species and growth forms: Fast growing branching corals tend to be more susceptible to bleaching and other stresses than slower growing massive corals. Some genotypes of a species will be more resistant to stress than other genotypes.	Do not focus on fast-growing forms but select a cross-section of species and growth forms that are relatively common and known to survive at your proposed rehabilitation site (or similar "reference" site). For individual species you should select a range of genotypes (i.e. try to obtain fragments from 30 or more donor colonies or from corals of opportunity scattered over a wide area that are likely to have come from diverse sources).
Coral species and growth forms that are not well adapted to the conditions at your rehabilitation site (assumed to be under some form of management control and with adequate water quality) will have low survival.	Base your selection of coral species and growth forms on what already survives (or survived in the past) at your rehabilitation site or at sites in similar environmental settings nearby ("reference" sites).
Life history and reproduction: If corals of the same species are attached too far apart from each other then fertilization rates of gametes may be inadequate (Allee effects).	Ensure you create adequate colony densities of each species in your transplant area. The density should be similar to that for the species at a similar healthy comparative area (reference site). (Don't place transplants of a species several tens of metres apart on a bare degraded reef.)

4. Social and political setting

Factors: Social and political setting in terms of national and local government units, relevant NGOs, coastal and watershed management initiatives; impact on stakeholder livelihoods; community acceptance (risk of interference); likely sustainability once initial funding comes to an end.

Information needs: You need a good understanding of the local decision-making processes and management influence(s) over the area under consideration and level of devolution of responsibilities/jurisdiction. A thorough ecological scoping exercise of potential sites will help in the assessment of the risk involved in the selection of sites. This will include an initial assessment of potential donor colonies and conclusions on the specific approach to restoration. You will need to be aware of the level of stakeholder understanding of the rehabilitation project and needs for training. Unpredictable confounding factors can be minimized by understanding the economic conditions of local communities most likely to be impacted by your project, including trends in local demographics. Post-project stakeholder behaviour can also affect your project so an understanding of the dynamics of local politics, social, and economic circumstances can help in building up stakeholder ownership and increase the likelihood of long-term community support.

Risk prediction	Mitigation or adaptive management response
If local anthropogenic impacts on the reefs are not under control then rehabilitation has a high risk of failure due to external impacts or local interference.	We strongly recommend that your restored site is subject to some form of protection (e.g. in a Marine Protected Area (MPA), marine reserve/park, *tabu* area, no-take area, etc.). By locating sites within existing broader protective regimes, a level of management and protection is maintained even when the project ceases to exist.
If social and political compromises have strongly influenced the selection of your rehabilitation site, perhaps overriding ecological criteria, then there is a high risk the project failing in the longer term.	If ecologically poor site selection is unavoidable you should make the potentially poor long-term outcomes clear to stakeholders at the planning stage. An alternative approach is to try to persuade stakeholders to include an additional site at a more ecologically favourable location.
Local policy and development decisions can impinge on a site after the transplantation work is complete, thereby changing the nature of governance and protection for your site. In subsistence community and traditional ownership settings, development or resource usage decisions can be unpredictable, but community acceptance can reduce the potential for change in a decision about a project's usefulness.	You should try to ensure that planned coastal developments are not likely to impinge on the restoration site over the expected lifetime of the project (including decisions on upstream locations), before you decide to go ahead with a project. Invest in strong community consultation at the project planning stage. Include a detailed and honest plan with timelines for outcomes that you ensure are well understood by the community. Your project should incorporate strong participation processes in all consultations, and present regular progress updates to the community throughout the life of the project.
Insufficient stakeholder understanding and training in basic biological requirements of reef rehabilitation can jeopardize the success of a project.	You should allocate sufficient resources and time to ensure you create high stakeholder awareness, acceptance, and understanding of the ecological background to your project. Stakeholder ownership and interest leads to vigilance towards threats.
Seemingly small changes in the social/political circumstances can alter the parameters around which a project was first designed and the ranking of potential threats.	Unpredictable changes in the social/political setting are risks that are hard to manage, but if you have strong and attentive management in the project you should be able to ensure that risks to the project are kept low by maintaining engagement with the community and local decision-makers.
Economic changes in the local community can affect the degree and type of resource use and put different pressures on resources. Such changes can cause an increase in unsustainable practices that may impinge on a rehabilitation site.	If you have designed a project with low overall risk, this allows for more flexibility to respond to unpredictable risks.

Successful recovery can produce new fish habitat, resulting in increased fish biomass (e.g. Ch. 8: Case study 10) and potentially a desire by local fishers to exploit this at the restored site. This may lead to damage to the rehabilitation site by fishing gear, anchors or spearfishing divers.	Ensure that long-term protection measures (e.g. sites within no-take zones) or fishery management regulations are in place and agreed with stakeholders at the planning stage. (It is easier to agree fishing regulations before significant biomass is achieved.) You can also maintain regular stakeholder consultations to reduce the temptation for damaging exploitation by encouraging local ownership of the site.
Often stakeholder interest in a project ceases after the funding finishes, resulting in the risk of neglect and damaging interference at the site.	Projects with long-term aims should include long-term involvement by project individuals as well as stakeholders to allow time for tangible benefits to develop. You need to foster stakeholder ownership during project life (particularly if tangible benefits such as fish are present) so that there are incentives to maintain the site and associated benefits post-funding.

5. Management issues

Factors: Personnel, capacity and training, relationship between project personnel and local government units, stakeholder engagement.

Information needs: For management of rehabilitation to be effective it should be adaptive (flexible and responsive to changing needs), thus you need to have a steady flow of information on progress of the rehabilitation project via a monitoring plan (Chapter 2). For a project to be sustainable in the long-term, it also usually needs to be firmly embedded in the local community with all key stakeholders involved from the planning stage (Chapter 2). The latter requires a flow of information from the project to stakeholders and decision-makers and the capacity for adaptive management responses to accommodate changes in stakeholder perceptions and aspirations, if appropriate.

Risk prediction	Mitigation or adaptive management response
A systematic monitoring and management regime that is too rigid can hinder timely adaptive responses to address unforeseen threats to a project and changing social, economic, political and environmental circumstances.	You should always include provisions for adaptive response measures in your project design and monitoring plan. You should distinguish between systematic monitoring for scientific reasons and for stakeholder and project donor reporting requirements, and monitoring for adaptive management reasons. Quick visual checks of the rehabilitation site on a weekly to monthly basis can both identify emerging threats and allow corrective responses to be taken in time.
A rehabilitation project that has relatively long timelines can become less efficient if key personnel leave, resulting in loss of critical understanding and focus. Key stakeholders, who hold the influence or decision-making power to maintain a project, may change over time resulting in decreased understanding and support for your project.	You can help new incumbents by providing appropriate detailed documentation of the project design, aims, timelines, and progress, including a good risk management plan, which will aid smooth staff turnover. You should ensure that personnel involved and stakeholders are aware of the essential project details so that local knowledge is retained. You should pro-actively engage with key stakeholders and local decision-makers so that they maintain support. When choosing your project staff, you could include an appropriate local person or group (e.g. NGO or Community Based Organisation) to manage stakeholder relations with the project.
Inadequate monitoring reduces the chances of successful adaptive management and potential for feedback to stakeholders.	At the planning stage, you need to assess the ability and training that will be required for staff to monitor key aspects of the project and provide feedback. You should regularly analyse and disseminate monitoring data to stakeholders, otherwise it is of little use.

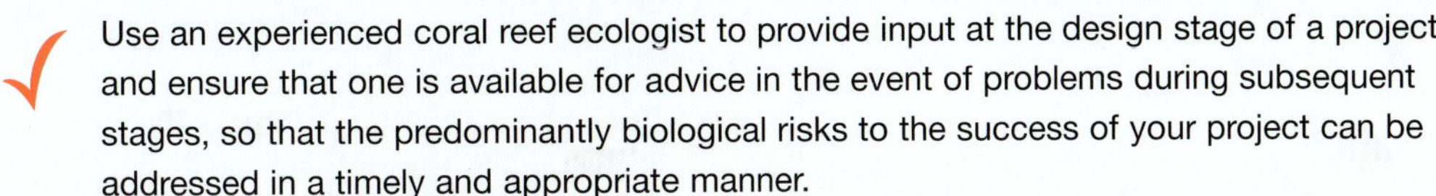

✓ Use an experienced coral reef ecologist to provide input at the design stage of a project and ensure that one is available for advice in the event of problems during subsequent stages, so that the predominantly biological risks to the success of your project can be addressed in a timely and appropriate manner.

✓ Make full use of the huge amount of useful information that is available free over the internet (see web links in reference sections of each chapter).

✓ Choose your rehabilitation site with great care and with full consideration of the ecological and logistic implications.

✓ Ensure that there is adequate management control of the rehabilitation site so that adverse human impacts (e.g. destructive fishing practices) will not jeopardise your project.

✓ Carefully research the optimum time of year for transplantation at your site and seek to avoid particularly stormy or warm times of year.

✓ To spread risk, replicate transplant sites whenever feasible, creating several well-separated patches of coral rather than a single large patch.

✓ Make sure you only transplant coral species or growth forms that are well-adapted to the conditions at your proposed rehabilitation site, using a "reference site" as a guide to what these are if need be.

✓ Within the constraints of the previous item, seek to use as wide a cross-section of common coral species and genotypes as you can, to increase the chance that some colonies will be resistant to any disturbances that may occur.

✓ Foster local support for your project by engaging fully with stakeholders, the community and local government units and use monitoring results to increase public awareness, report progress and maintain the project's media profile.

✓ Prepare for the unexpected by building into the project plan the capacity for monitoring and adaptive management in response to changing needs or setbacks.

Step 3. Assessing and prioritising the risks to your project

Once you have identified the main risks, you then need to assess them as to their potential severity and probability of occurrence. In some case these can be fairly easy to assess but in others they may be more or less impossible (e.g. in the case of the probability of a rare event occurring). Therefore, in the assessment process you may need to make educated guesses in order to prioritise the risks and develop a risk management plan. The person who is assigned to do this assessment will be critical to the whole design and risk management process. The pragmatic adoption of realistic aims will be equally important for the potential success of a project.

The fundamental difficulty in risk assessment is determining the probability of occurrence of a particular risk since statistical information is not available on past incidents that led to failure or significantly reduced successful outcomes. Furthermore, evaluating the severity of the consequences (impact) is often quite difficult also. You can attempt to quantify the risks by using the following approach: *Likelihood of occurrence* coupled with the predicted *consequences (impact of the event)* defines the severity of the *risk*, where a rank value can be assigned to each of the risk categories (Table 3.2). In the table, the predicted consequences of a risk are cross referenced to the predicted likelihood of occurrence to produce a potential risk ranking (ranging from 1 = near zero risk to 4 = high risk).

Where a number of specific risks are identified as likely to impinge on your project, a sum total of the risk rankings provides some guidance as to whether the project has a reasonable chance of success. The higher the sum of the risks, the higher the potential for failure or compromised project outcomes. The presence of several high risk factors would be particularly worrying. This exercise can be utilised in an iterative way, by altering the risk rankings according to the inclusion of additional provisions in the design phase that will mitigate certain identified risks (in the process, reducing the rank value of that risk), thereby improving the overall success potential of your project.

The final risk assessment matrix should be critically assessed as to whether it truly incorporates all of the most likely known and predicted risks. If there are several aims, then risks to one from a given factor may be greater than to another, in which case the more severe risk category should be chosen.

Table 3.2 A method to rank the severity of risks by incorporating the estimated likelihood of risk occurrence and predicted consequences of risk impact. H = High Risk (Rank value = 4), M = Medium Risk (Rank value = 3), L = Low Risk (Rank value = 2), NZ = Near Zero Risk (Rank value = 1).

Likelihood of risk impact	Consequences of risk occurrence			
	Severe	Moderate	Mild	Negligible
High	H	H	M/L	NZ
Medium	H	M	L	NZ
Low	M	M/L	L	NZ
Negligible	L	L	L	NZ

Step 4. Development of response options for mitigation of risks

Depending on the ranking of the most relevant and likely risks associated with a restoration project under each of the major aims, a list of appropriate responses can be developed. Suggested responses are outlined in Table 3.1. Note that specific risks can occur at various spatial and temporal scales so you most likely will have to address several risk factors in a given project. You may also need to develop site-specific responses that are not covered on this list.

The aim of mitigation is to ensure that perceived risks are minimised. If risks are in the medium or high category, a reassessment of how to minimise those risks may mean changing the initial project aims and design. Also, many risks might be associated with a selected site, so to minimise the risks, you may be required to look for an alternative site or to even abandon the project if viable alternatives are not available.

Table 3.1 is presented as a series of risk sources, the information you need to obtain in order to evaluate them, the specific risks associated with each source, and the appropriate management response to mitigate or minimise the threat from these risks. Many of the responses just involve careful planning and prior thought.

A local diver in Tuvalu inspects a coral transplant for unwanted predators and macroalgae on the skeleton prior to transporting it to a rehabilitation site (Ch. 8: Case study 2). The coral had feeding or disease scars which signalled that particular care was needed to ensure that no predators or disease were transferred to the rehabilitation patch. Alternatively, this colony may have had to be rejected (D. Fisk).

Step 5. Integration of earlier steps and development of a Risk Management Plan

At this stage, you may need to enter into an iterative process and reconsider your original aims or the suitability of your selected site in order to reduce the level of risk from the most significant of perceived threats.

Due to the uncertainties in our scientific knowledge, a key part of the Risk Management Plan may be to use monitoring to provide more information as the project proceeds and use this to guide adaptive management to correct problems arising, as well as to provide feedback to stakeholders. Thus monitoring is central to adaptive management and risk mitigation and not an optional extra. The key is flexibility and the capacity to respond to changing circumstances as indicated by the arrows in both directions between stages 4 and 5 in Figure 2.1 and those between these stages and the financial and human resources available.

Ultimately, all you can do is try to mitigate for likely risks and be in a position to respond to less likely ones. With careful project design and systematic consideration of the risk factors in Table 3.1 and their mitigation, you can reduce the chances of your rehabilitation project failing but you can never guarantee success.

References

1. Jaap, W.C. (2000) Coral reef restoration. *Ecological Engineering,* 15, 345-364.

2. Hudson, J.H., Anderson, J., Franklin E.C., Schittone, J. and Stratton, A. (2007) M/V *Wellwood* Coral Reef Restoration Monitoring Report, Monitoring Events 2004-2006. Florida Keys National Marine Sanctuary Monroe County, Florida. *Marine Sanctuaries Conservation Series NMSP-07-02.* U.S. Department of Commerce, National Oceanic and Atmospheric Administration, National Marine Sanctuary Program, Silver Spring, MD. 50 pp. [Download at: sanctuaries.noaa.gov/science/conservation/pdfs/wellwood2.pdf]

3. Precht, W.F., Aronson, R.B. and Swanson, D.W. (2001) Improving scientific decision making in the restoration of ship grounding sites on coral reefs. *Bulletin of Marine Science*, 69 (2), 1001-1012.

4. Kojis, B.L. and Quinn N.J. (2001) The importance of regional differences in hard coral recruitment rates for determining the need for coral restoration. *Bulletin of Marine Science*, 69 (2), 967-974.

5. Naughton, J. and Jokiel, P.L. (2001) Coral reef mitigation and restoration techniques employed in the Pacific islands. I. Overview. *OCEANS, 2001. MTS/IEEE Conference and Exhibition*, 1, 306-311.
[Download at: cramp.wcc.hawaii.edu/Downloads/Publications/CP_Coral_Reef_Mitigation_1.pdf]

6. Shaish, L., Levy, G., Katzir, G. and Rinkevich, B. (2010) Coral reef restoration (Bolinao, Philippines) in the face of frequent natural catastrophes. *Restoration Ecology*, [online: doi: 10.1111/j.1526-100X.2009.00647.x]
[Download at: www3.interscience.wiley.com/cgi-bin/fulltext/123320601/PDFSTART]

7. Reid, C., Marshall, J., Logan, D. and Kleine, D. (2009) *Coral Reefs and Climate Change. The Guide for Education and Awareness.* CoralWatch, University of Queensland, Brisbane. 256 pp.

8. Thom, R.M. (1997) System-development matrix for adaptive management of coastal ecosystem restoration projects. *Ecological Engineering*, 8, 219-232.

9. Green, A.L. and Bellwood, D.R. (2009) *Monitoring functional groups of herbivorous reef fishes as indicators of coral reef resilience – A practical guide for coral reef managers in the Asia-Pacific region*. IUCN working group on Climate Change and Coral Reefs. IUCN, Gland, Switzerland. 70 pp. ISBN 978-2-8317-1169-0.
[Download at: cmsdata.iucn.org/downloads/resilience_herbivorous_monitoring.pdf]

10. Jokiel, P.L. and Naughton, J. (2001) Coral reef mitigation and restoration techniques employed in the Pacific islands. II. Guidelines. *OCEANS, 2001. MTS/IEEE Conference and Exhibition*, 1, 313-316.
[Download at: cramp.wcc.hawaii.edu/Downloads/Publications/CP_Coral_Reef_Mitigation_2.pdf]

11. Guest, J.R., Dizon, R.M., Edwards, A.J., Franco, C. and Gomez, E.D. (2009) How quickly do fragments of coral 'self-attach' after transplantation? *Restoration Ecology*, 17 (3) [doi: 10.1111/j.1526-100X.2009.00562.x]

12. Harvell, C.D., Jordán-Dahlgren, E., Merkel, S., Rosenberg, E., Raymundo, L., Smith, G.W., Weil, E. and Willis, B.L. (2007) Coral disease, environmental drivers and the balance between coral and microbial associates. *Oceanography*, 20 (1), 58-81.
[Download at: www.tos.org/oceanography/issues/issue_archive/issue_pdfs/20_1/20.1_breaking_waves.pdf]

13. Bruno, J.F., Selig, E.R., Casey, K.S., Page, C.A., Willis, B.L., Harvell, C.D, Sweatman, H. and Melendy, A.M. (2007) Thermal stress and coral cover as drivers of coral disease outbreaks . *PLoS Biology*, 5 (6), e124, 1220-1227.
[Download at: www.plosbiology.org/article/info:doi/10.1371/journal.pbio.0050124]

14. Raymundo, L.J., Couch, C.S. and Harvell, C.D. (Eds) (2008) *Coral Disease Handbook. Guidelines for Assessment, Monitoring & Management*. Coral Reef Targeted Research and Capacity for Management Program, St Lucia, Australia. 121 pp. ISBN 978-1-921317-01-9

15. Bruckner, A.W. (2002). Priorities for effective management of coral diseases. *NOAA Technical Memorandum NMFS-OPR-22*. 54 pp.
[Download at: www.coral.noaa.gov/coral_disease/cdhc/man_priorities_coral_diseases.pdf]

16. McCook, L.J., Jompa, J. and Diaz-Pulido, G. (2001) Competition between corals and algae on coral reefs: a review of evidence and mechanisms. *Coral Reefs*, 19, 400-417.

17. McCook, L.J. (1999) Macroalgae, nutrients and phase shifts on coral reefs: scientific issues and management consequences for the Great Barrier Reef. *Coral Reefs*, 18, 357-367.

18. Mumby, P.J. and Steneck, R.S. (2008) Coral reef management and conservation in light of rapidly evolving ecological paradigms. *Trends in Ecology and Evolution*, 23 (10), 555-563.

19. Austin, J. and Atkinson, S. (2004) The design and testing of small, low-cost GPS-tracked surface drifters. *Estuaries*, 27 (6), 1026-1029.

20. Green, A., Lokani, P., Sheppard, S., Almany, J., Keu, S., Aitsi, J., Warku Karvon, J., Hamilton, R. and Lipsett-Moore, G. (2007) *Scientific Design of a Resilient Network of Marine Protected Areas. Kimbe Bay, West New Britain, Papua New Guinea*. TNC Pacific Island Countries Report No. 2/07. x + 60 pp.
[Download at: www.reefresilience.org/pdf/Kimbe_Complete_Report.pdf]

Chapter 4.

Constructing and managing nurseries for asexual rearing of corals

Shai Shafir, Alasdair Edwards, Buki Rinkevich, Lucia Bongiorni, Gideon Levy and Lee Shaish

4.1 Introduction

To initiate *active* rehabilitation of just a few hectares (1 ha = 100 m x 100 m = 10,000 m²) of badly degraded coral reef you are likely to need several tens of thousands of coral transplants. Without a nursery stage to rear coral "seedlings" from small fragments, the collateral damage to healthy reefs used for the collection of such material will generally be unacceptable. Thus a pre-requisite for reef restoration, at all but the smallest scale, is the establishment of coral nurseries that can supply large numbers (tens of thousands) of corals of a size that can survive and grow at the site to be rehabilitated. Such nurseries have been recently established successfully in the Caribbean, East Africa, Red Sea, Southeast Asia and the Pacific. Large-scale rehabilitation of reefs should thus be seen as a two-step process; firstly, the rearing in a nursery of coral "seedlings" to a size where they can be outplanted to the wild; secondly, transplantation of these to degraded reef areas[1]. This chapter provides information on the siting, construction, stocking, maintenance and costs of nurseries for rearing corals asexually. Rearing of coral larvae produced by sexual reproduction is discussed in Chapter 5.

Although it is a major cost in terms of staff time, careful and regular maintenance of the nursery promotes better survival and growth of healthy coral colonies and is crucial to cost-effective nursery operation. The success of a coral nursery depends largely upon how it is set up initially. The most successful nurseries studied to date are those located in mid-water at sites away from the natural reef, where corals are protected from predation by corallivorous organisms and interference by divers. However, for reasons of cost, shelter from storms, ease of access, construction and maintenance, one may need to site nurseries on the seabed in shallow and more accessible areas. In such nurseries corals may be at greater risk of predation, human interference, bleaching (if very shallow) and sedimentation but may nevertheless thrive sufficiently that the lower costs justify the somewhat higher losses. Cost-effective production of transplants in a nursery depends on careful management and adherence to protocols that produce healthy coral colonies suitable for transplantation in as short a time as feasible. Nursery design, substrates used for coral colony mariculture, realistic numbers and densities of colonies that can be maintained, duration of the nursery phase, growth rates and survival of farmed colonies are just some of the crucial issues which will be considered below.

There are several types of coral nursery that vary in structure, size and purpose. The major division is between *ex situ* nurseries, which are located on land, are expensive and largely for the specialist such as those culturing corals to supply the aquarium trade, and *in situ* nurseries that are located in the sea[2], are cheaper to construct and operate, and are the focus of this chapter. With guidance, the latter can be built and operated by NGOs and local communities. Many designs of *in situ* coral nursery have been tested around the world; we will discuss a few designs that use inexpensive and readily available materials, have been tested successfully at several sites worldwide and can be readily scaled-up to generate 10,000s of transplants annually. *In situ* coral nurseries can be on (but preferably are elevated 1–2 metres above) the seabed but fixed to the bottom ("fixed"), or they can be suspended in mid-water well-above the seabed ("floating"). To maximize survival we strongly recommend raising bottom-attached nurseries at least one metre above the seabed; this tends to reduce both predation by invertebrates and effects of sediment. Two main methods of construction will be presented for both types of nursery (fixed and floating). The first involves modular nurseries composed of trays constructed from plastic pipes and mesh on which coral "seedlings" are reared on pieces of substrate; the second involves rearing corals on ropes.

In this chapter we outline: 1) how to choose a site for a nursery, 2) how to construct nurseries for asexual rearing of corals, 3) issues to consider when stocking nurseries with corals, and 4) methods for maintenance of nurseries and their corals.

An ex situ coral nursery a few miles inland from the Mediterranean coast of Israel. Note the expensive aquarium equipment needed to maintain the corals (S. Shafir).

An in situ floating nursery in clear water in the northern Red Sea. Note the use of mesh trays to hold coral fragments mounted on plastic pins as in the ex situ nursery (S. Shafir).

4.2 Selecting a site for a nursery

When choosing a site to establish a nursery, you should consider the following points: water quality, shelter, accessibility, and tidal range. Above all the nursery site should be appropriate for rearing corals that will survive at the site where you intend to transplant them; thus conditions at the nursery (depth, water temperature, salinity, sedimentation, etc.) should be reasonably similar to the transplant site. For example, if you are trying to rehabilitate a degraded reef patch at 2 m depth in a lagoon, you would not set up your nursery at 10 m depth on the reef slope, and vice versa.

You need to select a site which will allow as little maintenance as possible and good survival. Choose a site with good water quality not subject to freshwater or sediment laden runoff from land during rainstorms. Ensure there are no nearby sources of human pollution (e.g., sewage, excessive aquaculture pond effluent) that may affect the site. Although a certain level of nutrients§ can augment coral growth (shortening the time the corals need to be in the nursery), elevated nutrient levels and suspended particulate matter can be harmful and encourage algae and encrusting sponges, tunicates, bryozoans, barnacles and molluscs which then smother and kill corals in the nursery. Seek a site where water clarity allows adequate light penetration for good coral growth at the depth where the corals will be cultured. The depth of the nursery should be similar to the depth of the site where you intend to transplant.

Sedimentation and colonisation of the nursery structure by fouling organisms are two main killers of coral "seedlings". Regular maintenance (section 4.5) is required to remove both fouling organisms and accumulation of silt. Look for a site with as little sedimentation as possible.

Corals can bleach as a response to stresses such as sudden changes in water temperature or salinity, and can die if such stresses are prolonged. Thus try to select a site with sufficient water exchange that temperature and salinity are fairly stable. Lagoons or closed bays may offer shelter from storms but during neap tides at the warmest time of year may be subject to excessive warming, particularly on sunny days, such that corals bleach (Figure 4.1), and during heavy rains may be subject to low salinity from surface run-off or groundwater upwelling.

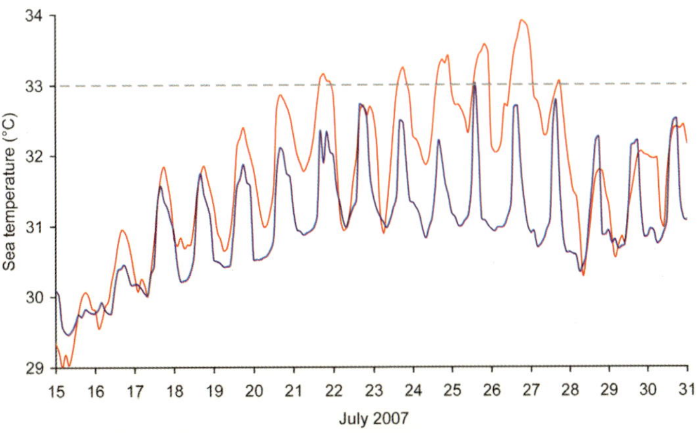

Figure 4.1. Comparison of temperatures at two transplant sites about 2 km apart in the same lagoon during part of one of two warmest months of the year for this location (north-west Luzon in the Philippines). The red line shows temperatures at a transplant site in the middle of the lagoon at about 2.7 m depth and the blue line the temperature at 4.4 m depth at a second site near a channel leading to the open sea. The graph shows the daily fluctuations in water temperature (sometimes exceeding 2°C) to which the transplanted corals were subjected and how the period of high temperature tended to last longer each day at the more sheltered site. Sea temperatures above 33°C cause extensive coral bleaching at this location and the small temperature differences between the two sites meant that some coral species that bleached and suffered high (68-100%) mortality at the mid-lagoon site were more or less unaffected at the other site. This illustrates the importance of carefully selecting sites for both nurseries and transplantation.

§ At an oligotrophic site in the Gulf of Aqaba (ambient monthly average nutrient levels: 0.075 µmol/L nitrite, 0.264 µmol/L nitrate, 0.045 µmol/L orthophosphate, 0.057 µmol/L ammonia), corals grew 3 times faster adjacent to a commercial fish farm (monthly average nutrient levels: 0.095 µmol/L nitrite, 0.385 µmol/L nitrate, 0.123 µmol/L orthophosphate and 1.016 µmol/L ammonia)[3].

Nursery sites should be chosen with careful thought about how deep the corals being farmed will be at lowest low tide. A minimum of about two metres of water cover during lowest tides is recommended even in relatively sheltered areas where waves are of less concern. Although good water exchange is beneficial, avoid areas that may be subject to strong tidal currents that can damage the nursery structure.

Accessibility

Choosing the site for a nursery always involves trade-offs. A fixed nursery at a site close inshore in a reasonably sheltered area, which can be reached without using a boat and is shallow enough to be maintained by snorkelers, needs minimum logistical support. A floating nursery at a site further offshore which requires a boat for access and SCUBA to dive down to maintain the farmed corals needs much greater logistical support. The former may potentially be cheaper to operate but is likely to more prone to human impacts and attack by coral predators. The latter may be more prone to storm damage, even though floating nurseries can be lowered several metres in the water column when storms approach (and have thus survived hurricanes in Jamaica). Try to find suitable areas that are sheltered from storms and ocean swells and remember that areas that appear sheltered during one monsoon season may be unworkable during much of the other monsoon season. The knowledge of local people, particularly fishers, will be invaluable in site selection.

4.3 Nursery construction

Modular plastic pipe and mesh tray nurseries have now been tested successfully in a range of environments at sites in several countries around the world. Rope nurseries are cheaper to construct and operate but deployment strategies for rope-reared colonies are still being tested and, until there is evidence that rope-reared colonies can be deployed successfully, these nurseries should be regarded as experimental. Both types of nurseries can be constructed from inexpensive materials that are readily obtainable from hardware and plumbing stores in most parts of the world. Some tested designs are described below with the focus on how to construct floating and fixed nurseries to yield 10,000 transplants per year as this is the order of numbers likely to be needed to rehabilitate reefs at a scale of hectares. Smaller nurseries that are easier to construct and manage, and can generate several hundreds to several thousands of transplants per year are also described. All the nurseries are modular and therefore readily scalable to the needs of a particular rehabilitation project.

When establishing coral fragments in a nursery, you need to be thinking ahead to how you will deploy the colonies (once

Left: *A coral "nubbin" (small fragment) of* Pocillopora damicornis *mounted on a plastic pin with a drop of superglue. Scale on ruler is in mm (S. Shafir).* Right: *An* Acropora variabilis *fragment that has self-attached and grown over the plastic pin head after c. 4 months in a nursery but is primarily growing vertically (D. Gada).*

grown) on the degraded reef (see Chapter 2), as this is the key to cost-effective transplantation. The quicker colonies can be securely attached to a degraded reef, the more cost-effective the method. Coral fragments also need to be attached in the nursery and if they can be grown in the nursery on a substrate that can be readily fixed to the reef, this can greatly improve cost-effectiveness. With this in mind, we provide a brief overview of rearing substrates (section 4.4) and the deployment of transplants in the next two paragraphs, dealing firstly with branching and then with massive and encrusting growth forms.

For modular tray nurseries, you can wedge branching coral fragments in large plastic wall-plugs (e.g. those for 8-12 mm diameter drill-bits), plastic push-plugs, or insert them into short lengths of plastic tubing or hose-pipe (with diameter chosen to hold the corals securely), or cement them with superglue to plastic pins[§]. In each case the plastic substrate is fixed in the nursery by inserting it in round or square-holed plastic mesh. The mesh size must be carefully chosen so that the substrate is held securely by the mesh. The coral base generally grows over the substrate forming a secure bond. At transplantation an underwater compressed air drill is used to make holes in the degraded reef of the correct size for the wall-plug, push-plug, tubing or plastic pin and the cultured coral can just be slotted into the reef. If the fit is poor then a little epoxy putty can be used to ensure good attachment. If the coral rock is not too hard, then you can use a hand-auger to make holes for the plugs or plastic pins (Chapter 6).

For rearing, you can attach fragments of massive and encrusting coral species with superglue directly to plastic mesh, to plastic pins inserted in the mesh, or to small

---

§ These have been used extensively in Red Sea nurseries. These have a 2 cm diameter head and a 9-cm long, tapered pin (0.3-0.6 cm wide). These are a waste product from plastic injection moulding in plastics factories.

plastic squares, calcareous or ceramic substrates. The latter flat substrates should have two holes drilled in them so that they can be tied to the mesh trays with cable ties, fishing line or plastic coated wire. The substrate can influence the growth form of the colony; a wide flat surface allows the coral fragment to encrust laterally whereas the narrow head of a plastic pin promotes upward growth. Once the encrusting or massive corals have grown to a suitable size, mesh can be attached to the reef using masonry nails or large barbed fence staples (or similar building products), as can other substrates if they are pre-drilled with holes. Plastic pins can be attached as indicated above. Otherwise, you can attach substrates to patches of reef (scraped or wire-brushed clean) with epoxy putty.

For rope nurseries, the substrate is the rope itself. Once colonies have grown to the desired size, the rope (with attached colonies) can be fixed to the reef with stakes or masonry nails. Pilot experiments, which are testing the efficacy of this approach, are still on-going so although we know that rope nurseries are very cost-effective for rearing corals, we are still uncertain about how to deploy the products successfully.

Fixed nurseries

Designs for two types of fixed nursery are given: firstly, for a fixed modular tray nursery; secondly, for a fixed rope nursery. The designs given are for guidance and are not meant to be proscriptive. They can no doubt be modified and improved. Many slightly different versions of the fixed modular tray nursery have been built by groups working in different countries and all seem to have worked satisfactorily.

Fixed modular tray nurseries

For a modular tray fixed nursery we suggest using trays (say 30 cm x 50 cm, or 60 cm x 80 cm) connected by cable-ties or cheaper but longer lasting mono-filament fishing line to rectangular frame-tables (say 1.2 m x 3.5 m to hold 20 small trays, or 1.4 m x 4.3 m to hold 10 large trays). The frame-table area should allow a few centimetres gap between trays for ease of working. Trays can be constructed of 1.25–1.6 cm diameter PVC pipes (stronger 2 cm diameter pipe is needed for larger trays or more exposed sites) made with plastic mesh attached by cable-ties (or monofilament line) to the pipe rectangles[4]. If corals will need to be transported significant distances in containers of seawater to the transplant site then the size of the trays should be such that they will fit in available containers. Thus the smaller 30 cm x 50 cm trays might be appropriate. It is difficult to find seawater containers large enough for the larger trays, thus these are appropriate where the nursery site is close enough to the transplant site so that trays can be transferred underwater by divers. (1 m x 1 m trays have been trialled but when full of grown corals they are heavy, unwieldy and difficult to handle underwater.) Examples of calculations of number of trays and frame-tables needed for rearing set numbers of corals at two spacings are given in Box 4.1.

The frame-tables can be made of approx. 3 cm diameter PVC pipe with legs of 3 cm wide angle-iron hammered 60-100 cm into the seabed (Figure 4.2). It is critical that you anchor the legs of the frame-tables securely. Attachment of fixed nurseries in soft sediment lagoon areas is a relatively easy task but becomes more complicated if the nursery has to be deployed on a reef slope. In this case the length of the angle-iron legs should be planned to ensure the frame-table is horizontal with down-slope legs longer than up-slope legs. We recommend that you brace the legs with connecting diagonal and horizontal bars of angle-iron. If the chosen nursery site is sometimes subject to wave action or strong currents, we suggest that you embed the legs of the nursery in concrete blocks to provide additional stability. PVC pipes for table-frames and trays generally last for several years and are readily obtainable from hardware stores around the world. You can drill a few small holes into the pipes to allow them to fill with water and reduce buoyancy, but beware of weakening the pipes. You may be tempted to use alternative cheap and available materials, such as bamboo, but these are likely to degrade quickly and may compromise the nursery structure leading to loss of corals. In our experience this is a false economy.

Left: *A fixed modular tray nursery on the seabed in the murky, sediment-laden waters of Singapore (A. Seow).* Right: *A different design with four 60 cm x 80 cm rearing trays on an angle iron frame in a lagoon in Philippines. The corals here are being reared for a scientific experiment rather than for restoration and are held in place by pegs (D. dela Cruz).*

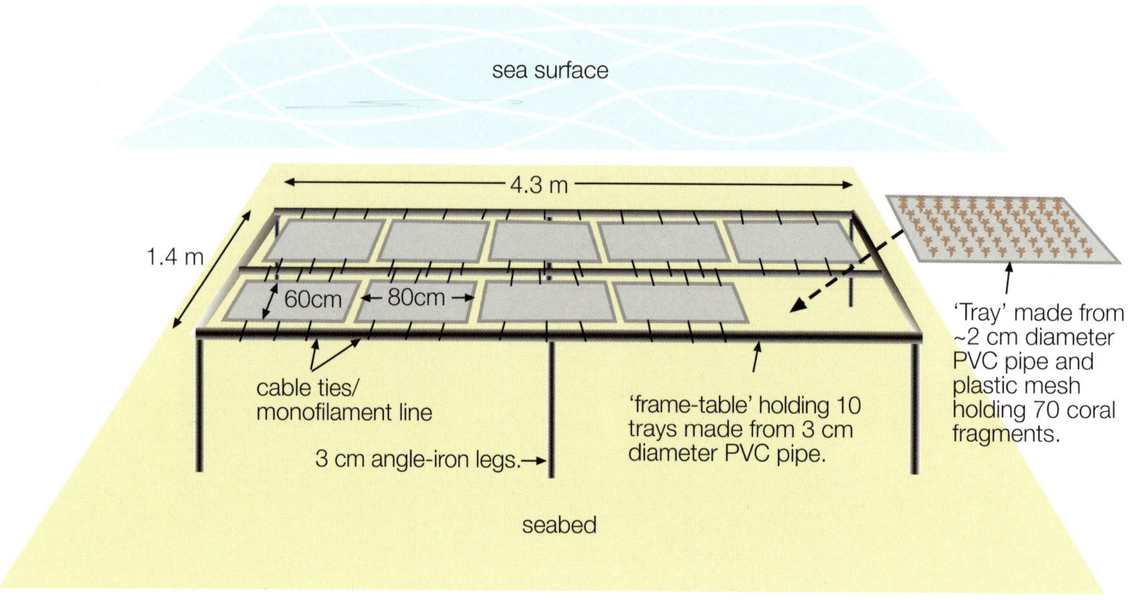

Figure 4.2. Design for a fixed modular tray nursery with the capacity to rear up to about 700 coral fragments - based on Shaish et al. (2008)[4].

Box 4.1 Calculating numbers of trays and frame-tables needed to rear target numbers of fragments at different spacings.

A 30 cm x 50 cm tray can hold about 15 corals spaced at 10 cm, or c. 45–55 corals spaced at 5 cm. A 60 cm x 80 cm tray can hold c. 40 corals spaced at 10 cm, or c. 135–145 corals spaced at 5 cm. Spacing of corals will depend on their growth rates and initial size. Fast growing branching species (e.g. *Acropora, Montipora, Pocillopora*) should be spaced further apart than slow growing forms (e.g. *Pavona, Porites, Heliopora*). During maintenance visits you can space out coral fragments if individuals are coming into contact but such work carries a cost, so we recommend adjusting spacing at the start to minimise maintenance needs.

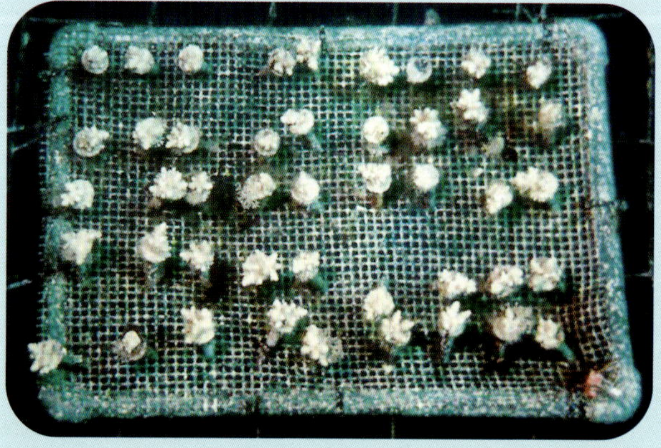

Coral rearing tray made of plastic mesh stretched over a 30 cm x 50 cm rectangle of 2 cm diameter plastic pipe (as used in plumbing). The mesh is held in place by cable-ties. This tray carries 5 rows of 9 Acropora fragments which have been attached to plastic pins with superglue a few months previously. The corals are spaced about 5 cm apart in the tray which is itself secured by cable-ties to a net which forms the 1.2 m wide working area of the mid-water floating nursery (S. Shafir).

To produce 10,000 transplants per year using the small (30 cm x 50 cm) trays and allowing for a 20% loss of corals in one year, you would need 800 trays (spacing 10 cm) or c. 240 trays (spacing 5 cm), equivalent to a table area of 120 m² or 36 m² respectively. To accommodate these numbers of trays, you would require 40 or 12 small (1.2 m x 3.5 m) frame-tables respectively. Using the large (60 cm x 80 cm) trays and again allowing for a 20% loss due to mortality, you would need 300 trays (spacing 10 cm) or 86 trays (spacing 5 cm). To accommodate these numbers of large trays, you would require 30 (spacing 10 cm) or 9 (spacing 5 cm) large (1.4 m x 4.3 m) tables respectively. Such calculations need to be carried out during the planning process (see Chapter 2).

Fixed rope nurseries

The need for restoration to be cost-effective has led to experimentation with even simpler structures that can be cheaply built from easily procured material, with a minimum of technical knowledge. Various researchers around the world have independently experimented with rearing or deploying corals on monofilament line and polythene string in low energy areas. This has led to the development of 'rope nurseries'. Prototype rope nurseries have been constructed and tested in the Indo-Pacific and the Caribbean and have delivered promising results. One of the largest costs in reef restoration is transplanting nursery-reared corals to the site being rehabilitated (see Chapter 7). Any method that can reduce this cost can dramatically reduce the overall costs of rehabilitation. An advantage of rope nurseries is that the nursery substrate (natural fibre or plastic [e.g., polypropylene, nylon] rope) may also serve as the means of attachment to the reef substrate at transplantation. Thus the ropes, with the developed coral colonies, can be transplanted 'as is', by anchoring the line with masonry nails or metal stakes to either, hard or soft substrates depending on the coral species being reared (Chapter 6). Not all coral species are amenable for rearing in rope nurseries (e.g. massive and slow growing species are less suitable). Branching corals and some encrusting forms appear to grow well. Rope nurseries are not only cheap to build compared to other types, but also need relatively low maintenance and allow fast deployment of transplants. Although trial deployments of a few species, still attached to the ropes on which they were grown, have shown promise, good long-term survival (e.g. over several years) of such transplants has <u>not</u> yet been demonstrated.

Like modular tray nurseries, rope nurseries may be constructed as floating or fixed. You can make a simple fixed rope nursery that will mariculture approximately 1000 corals from six 2.5–3.0 m long angle-iron bars hammered vertically into sandy substrate at 5 m intervals to form a framework from which ropes can be suspended (Figure 4.3). Each pair of verticals is connected at the top by a 1.5 m length of angle-iron to make a 10 m x 1.5 m frame over which natural fibre or plastic ropes (with coral fragments inserted) can be suspended. To strengthen and stabilize the structure, you can attach the tops of the vertical angle-iron bars at each corner by fishing line or plastic twine/rope to short lengths of angle-iron or rebar hammered into the sand (acting like guy-ropes for a tent). Coral fragments are inserted at c. 10–15 cm intervals into 6–8mm diameter natural fibre or plastic rope by temporarily untwisting the rope every 10–15 cm and sliding the fragments between the strands, allowing the twist of the rope to hold the fragment in place. You can adjust spacing between corals according to the rate of growth of the species and how long you intend to maintain the corals in the rope nursery prior to transplantation. The ropes with the inserted coral fragments are stretched between the framework by tying them to the angle-iron horizontals, allowing a spacing of 15 cm between adjacent ropes. A 10 m x 1.5 m frame allows ten 10-m rearing ropes to be accommodated comfortably. The material costs for such a rope nursery module are about US$100. Ten such structures could generate 10,000 corals per year. Deployment of corals reared in rope nurseries is discussed further in Chapter 6.

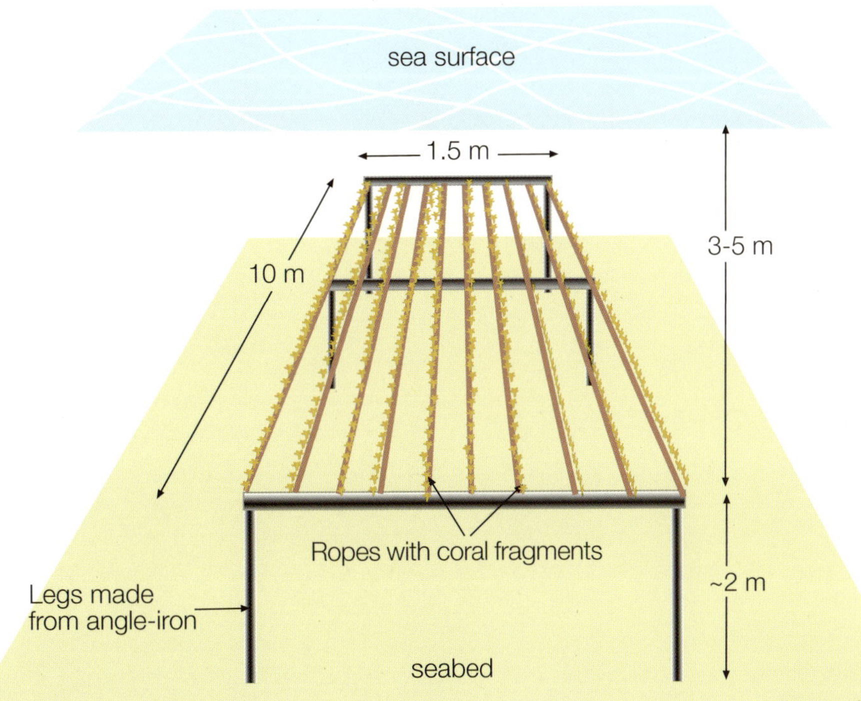

Figure 4.3. Design for a fixed rope nursery where 10-m lengths of 6–8 mm diameter plastic or natural fibre rope with coral fragments inserted every 10–15 cm are reared on a frame made of angle-iron. (Based on Levy, G., Shaish, L., Haim, A. and Rinkevich, B. (2009) Mid-water rope nursery–Testing design and performance of a novel reef restoration instrument. *Ecological Engineering,* doi: 10.1016/j.ecoleng.2009.12.003).

Floating nurseries

A floating nursery can be constructed in a range of habitats from lagoons a few metres deep to offshore (>20 m deep) blue-water areas away from the reef. The depth of the culture trays and type of floating nursery decided on will depend on the nature of the site selected for rehabilitation. We describe designs for three types of floating nursery: firstly, one for a large (10 m x 10 m) open-water nursery with the capacity for rearing 10,000 corals per year; secondly, one for a simple, multi-use, mini-floating nursery (1.2 m x 1 m), that can hold 400–600 small coral fragments at a time; and thirdly, one for an easy-to-construct, lagoonal floating nursery that can culture about 1300–1800 corals. About eight of these latter nurseries would allow 10,000 corals a year to be generated. Like fixed nurseries, floating nurseries need to be at sites where human activities can be controlled. Even in marine protected areas (MPAs), there can be problems, for example, a lagoonal floating nursery in a Zanzibar MPA was severely damaged in 2009 after being snagged by a trawler fishing illegally at night.

Open-water floating nursery

One of the advantages of constructing a mid-water nursery well above the seabed and well away from reefs is its isolation from coral predators, disease vectors and sedimentation effects. We recommend that mid-water nurseries are sited in a depth of at least 20 m over a sandy seabed. A hollow-square design that has been tested in several countries and allows divers to maintain the farmed corals easily is described[5]. Dimensions are for a 42 m² tray nursery which could produce 10,000 corals per year. The outer sides of the horizontal hollow-square are 10 m x 10 m and the inner sides are 7.6 m x 7.6 m (Figure 4.4). This forms a 1.2 m wide mariculture working area around the rim of the nursery that divers can access from both inside and outside without risk of accidentally damaging corals with their fins. The structure can be built from 6–8 cm diameter PVC pipe and the working area is covered by netting (5 cm x 5 cm and 10 cm x 10 cm netting have proved effective) to support trays of corals. The advantage of finer netting is that it may catch corals that are accidentally dislodged from the trays; the disadvantage is that it presents a greater surface area for fouling organisms which add weight to the nursery and need to be cleaned off periodically. You can reduce the latter problem by stringing 4–6 mm diameter ropes across the 1.2 m wide working area at 20 cm intervals for 30 cm x 50 cm trays (or at 50 cm intervals for 60 cm x 80 cm trays) instead to support the trays. As for fixed nurseries, you should choose a tray size such that grown corals can readily be transported to the transplant site in the trays. For transport over distances that require the corals to be submerged in seawater, it is easier to find large enough containers for the smaller trays.

View looking upwards from beneath a 10 m x 10 m floating nursery capable of farming 10,000 coral fragments a year (S. Shafir).

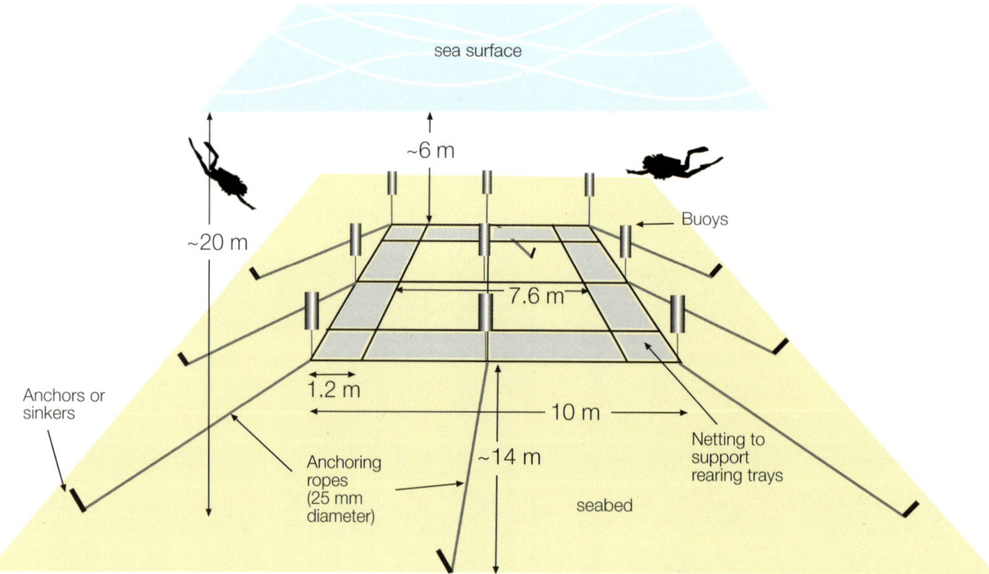

Figure 4.4. Design for a mid-water floating nursery with capacity for rearing 10,000 corals per year – based on Shafir and Rinkevich (2008)[6]. Depths will vary from site to site and depth of the nursery should be optimised for the site being rehabilitated.

Anchorage of floating nurseries can be difficult. The seabed anchors must hold the nursery fixed in place despite waves, the fall and rise of the tide and currents, and counteract the buoyancy supporting the coral trays. A floating nursery of this size would need to be attached to the sea bottom by eight anchors or sinkers; one from each of the four corners and one from the middle of each side. You can make these from 50-m lengths of ship anchor chains (each weighing about one tonne), or 2-m long, heavy 5–7 cm diameter galvanized metal pipes hammered diagonally into the seabed. (Pipes can also be drilled into the seabed using a water pump which can deliver at least 4–6 bars (400–600 kPa) pressure, placed on a boat with long air-hoses to the bottom.) Sinkers for this size of nursery can also be made of concrete blocks weighing at least one tonne or of oil drums filled with concrete. A dual system comprised of both sinkers and anchoring pipes hammered into the seabed should ensure a stable anchorage for the nursery during storms.

A 2-m long, heavy 5–7 cm diameter galvanized metal pipe hammered diagonally into the seabed being used as an anchor for a floating nursery above a sandy bottom (S. Shafir).

Buoys are needed to keep the nursery tensioned, more or less neutrally buoyant, and level at the planned depth. Here a 40 litre plastic container filled with air and attached via a cradle of rope (to reduce stress at attachment point) is shown being used as a buoy. Three small spherical buoys and a large white buoy (60 litre) are also visible. As the corals grow and add weight to the nursery, buoyancy needs to be increased (S. Shafir).

Nine buoys (approx. 40 litre) directly attached to the nursery structure are used to suspend the nursery in the water column and maintain the structure in tension (Figure 4.4). You can adjust buoyancy by adding air to the buoys (or attaching additional buoys) as the weight of the nursery increases during the mariculture period. If necessary, buoys can be submerged and maintained about 1 m above the chosen nursery depth to avoid collision with boats. Using submerged buoys also reduces the chance of attracting human interference. Anchor ropes (25 mm diameter is appropriate for this size of nursery) should have sufficient extra length to allow the depth of the nursery to be adjusted as required.

Coral fragments attached to plastic or other substrates are placed on mariculture trays covered with a c. 0.5 cm plastic mesh. If using 50 cm x 30 cm rearing trays, each tray can hold about 50 corals at the normal initial spacings used. For the 42 m² working area design described over 250 such trays can be maintained, allowing mariculture of about 10,000–12,500 corals depending on the size and colony growth form.

4

Mini-floating nursery

A mini-floating nursery[6] can be useful if a) you require relatively small numbers (a few hundred) of nursery reared transplants, b) you have very limited source material and thus need to rear from small nubbins (c. 0.5 cm²) which need extra early care and are initially grown at high densities before transfer to the main nursery, c) you need gradually to photo-acclimate batches of transplants from nursery depth to transplant site depth, or d) you wish to test nursery performance at several sites before deciding which is the right site for a large nursery.

When rearing from a very small size, more cleaning is required and mini-floating nurseries can easily be raised in the water column to 0.5–1 m depth for a short period (preferably on cloudy days or early morning or late afternoon to minimise light stress) to allow cleaning by snorkellers. After cleaning, you should immediately return the corals to normal rearing depth. Once corals have reached 3–5 cm diameter, you can transfer them to the main nursery.

Although we strongly recommend rearing corals at as close to the depth at which they will be transplanted as possible, this may not always be practical for logistical reasons. In such cases, we recommend that you photo-acclimate batches of corals prior to transplantation by transferring from the main nursery to a mini-floating nursery and then acclimating them in stages (e.g. from 8 m to 6 m depth for one week, then 6 m to 4 m depth for one week for corals reared at 8 m depth which are to be transplanted at 4 m depth). Photo-acclimation should be done with caution particularly to shallower depths (< 5 m). We recommend that normally transplantation should not be undertaken at sites < 2 m below lowest low tide and that the depth-change between rearing and transplant site should not be more than 50% of the rearing depth.

A mini-floating nursery can be constructed from a c. 1.2 m x 1 m plastic pallet (for an example, see www.nelsoncompany.com/prodplasticpallets.cfm) with two 1–2 litre buoys attached to the underside of <u>each side</u> of the pallet (Figure 4.5). A 2-m length of 8 mm diameter plastic rope is tied to each corner of the pallet and these are tied to a stainless steel ring below the centre of the pallet. A vertical anchoring rope with a 20–40 kg concrete sinker (depending on exposure to wave motion) at one end is passed through the ring and the free end tied off to the anchoring rope a metre or two below the ring. This rope is used to adjust the depth of the nursery and so the length of the free end should be at least equal to the depth of the nursery so that the mini-floating nursery can be raised to just below the sea surface for cleaning. As an alternative to a 20–40 kg sinker, you can use a 1-m long, 5–7 cm diameter galvanized metal pipe hammered into the substrate.

Side-view of a mini-floating nursery made from a c. 1.2 m x 1 m plastic pallet supported by 1–2 litre buoys (plastic bottles) tied to the underside and with trays of corals growing on top (S. Shafir).

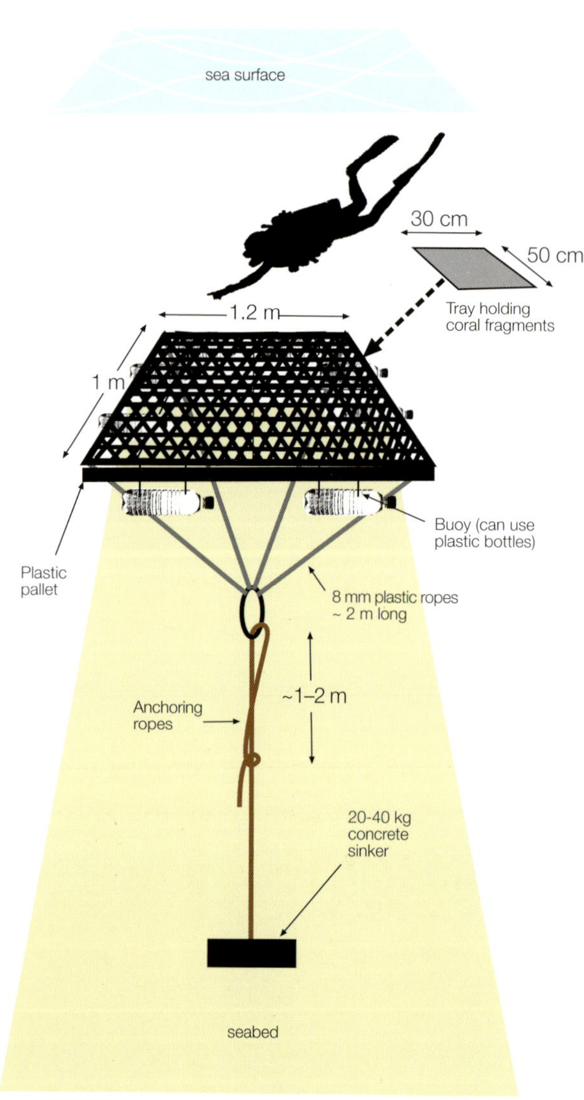

Figure 4.5. Mini-floating nursery design – based on Shafir and Rinkevich (2008).

Plastic bottles may be used as buoys. Depth adjustment of the mini-floating nursery is achieved by releasing or pulling the rope through the central ring and tying it when the desired depth is achieved. Such a nursery can be made for about US$30–50.

Lagoonal floating nursery

In lagoonal areas with at least 5–10 m water depth, medium-sized floating nurseries have been successfully trialled. These offer most of the benefits of an open-water floating nursery but are easier to construct and maintain. We provide an example design for a 1.2 m x 5 m nursery which could support 32 mesh trays (30 cm x 50 cm) carrying approximately 1300–1800 corals. (Thus about eight of these nurseries would allow 10,000 corals a year to be generated.) The rectangular 5 m x 1.2 m frame is made of 6 cm diameter PVC pipe. We recommend having a cross-pipe approximately every 2 m–2.5 m to strengthen the frame. A net (e.g. 10 cm x 10 cm mesh size, but confiscated fishing net has been used reasonably successfully) is stretched across the frame being tied to the pipe with cable ties or monofilament line (or cross ropes can be strung between the long sides at intervals suitable for the mesh trays being used). Size of frames should be adjusted to the characteristics of the site (e.g. larger frames can be deployed in large sheltered sites than in small or more exposed ones) and according to the number and size of cultured colonies required. We recommend small mesh trays (30 cm x 50 cm) for rearing the corals when several coral species are maricultured, with each tray holding a single species (or a single genotype) to reduce the chance of negative interactions between adjacent cultured fragments. Small trays facilitate management. You can secure the mesh trays to the pipe frame and net (or cross ropes) by cable-ties or mono-filament fishing line (which is cheaper and lasts longer but is harder to tie securely).

The lagoonal floating nursery at Chumbe Island, Zanzibar. Here large mesh trays (1 m x 1 m) are being used to culture Porites cylindrica *fragments inserted into short lengths of hosepipe. We recommend using smaller trays for rearing. The frame-table is about 5 m above the seabed and 5 m below the surface (S. Shafir).*

Corals of several species growing in a rope nursery near Bolinao in the Philippines. Each colony started as small fragment (G. Levy).

As well as anchoring the structure to the seabed, you should buoy it at the four corners and at about every 2.5 m along the long sides (to avoid deformation of the structure) using plastic ropes to connect the buoys (5–10 litre plastic containers) to the frame. (If using water or cooking oil containers, the strain at the attachment point (neck or handle of the container) can be reduced by passing the rope over the base of the container (now floating uppermost) to form a cradle and spread the load.) Supporting buoys can float on the surface but it is recommended to use buoys submerged at least 2–3 m, as these are less subject to wave movements and collision with boats, and less likely to attract people who might interfere with the nursery. You can estimate the weight of the nursery in order to calculate the buoyancy needed. One way is to measure the buoyant weight of one colony and multiply up to calculate the total weight of the nursery. Another way is the empirical approach which is to fill the plastic containers gradually with air (or add buoys) until the nursery floats. During the nursery period, you should adjust the amount of buoyancy as the weight of the nursery increases with coral growth. The aim is to keep the nursery slightly positively buoyant so that there is some tension in the structure. For larger (over 4 m²) nurseries, we recommend that you use at least four 25–40 kg sinkers or anchors (e.g. 1-m long, 5–7 cm diameter galvanized metal pipe hammered into the substrate), each attached to one corner of the rectangular frame by a vertical rope in order to maintain the structure in tension.

Floating rope nursery

A floating rope nursery with the potential to culture around 10,000 fragments has been trialled in Philippines. A modified design based on this is presented here. This involves a series of seven parallel, horizontal 8-m lengths of 8 cm diameter plastic pipe buoyed at each end with subsurface 5–10 litre buoys or plastic containers and anchored to the seabed by vertical ropes attached either to 100 kg sinkers or 2-m lengths of rebar, angle-iron or heavy 5–7 cm diameter galvanized metal pipes hammered diagonally into the seabed (Figure 4.6). The horizontal plastic pipes are spaced at 5 m intervals and serve to support about forty 30-m lengths of 6–8 mm natural fibre or plastic rope spaced approximately 20 cm apart. You insert coral fragments at c. 10–15 cm intervals in the rope by temporarily untwisting the rope every 10–15 cm and sliding the fragments between the strands, allowing the twist of the rope to hold the fragment in place. You can adjust the spacing between fragments according to the rate of growth of the species and how long you intend to maintain the corals in the rope nursery prior to transplantation. The ropes with the inserted coral fragments are stretched across the seven horizontal pipes, allowing a spacing of c. 20 cm between adjacent ropes, which are secured to each pipe by cable-ties. This design was initially built using local bamboo for the cross-pipes but the material did not last well and is not recommended.

Two c. 25 kg concrete sinkers anchoring two corners of a 1.2 m x 5 m lagoonal floating nursery at Chumbe Island, Zanzibar (J. Guest).

Fouling organisms on rearing trays in a Red Sea coral nursery. The sea anemones Aiptasia pulchella *and* Boloceroides mcmurrichii *are particularly in evidence (S. Shafir).*

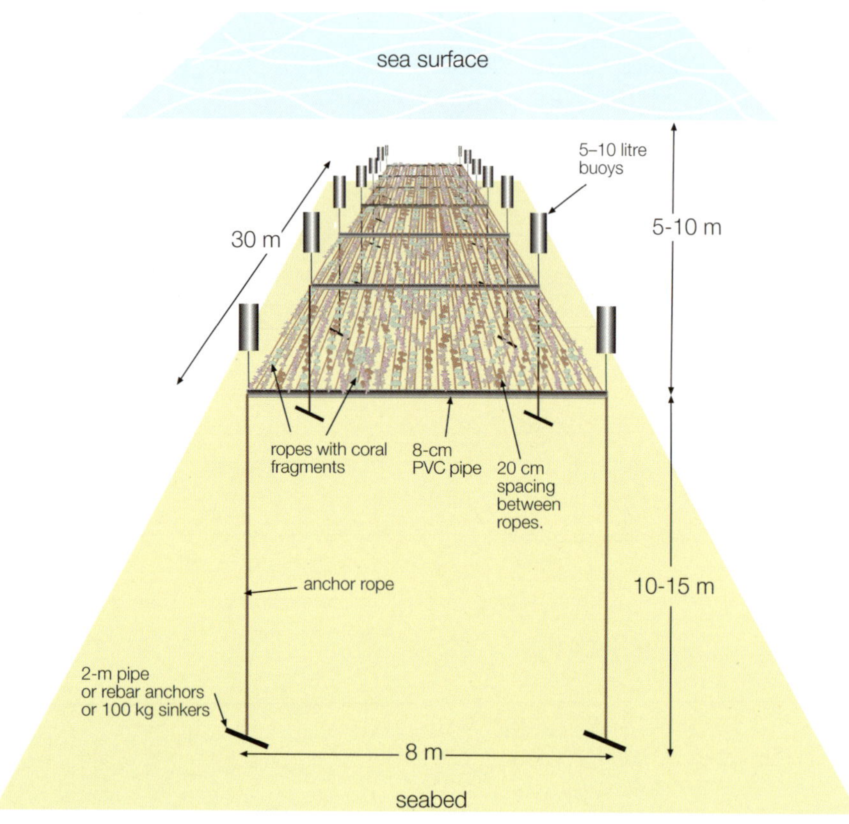

Figure 4.6. Design for a large floating rope nursery with capacity to rear about 10,000 coral fragments.

Fouling issues and use of anti-fouling paints

Fouling of the nursery structure, that is, growth on it of algae and invertebrates such as sea-squirts, sponges, bivalves, barnacles, etc. is generally a significant issue even at the best-sited nurseries. The problem is greater in modular tray nurseries, because of the huge surface area of mesh that can become fouled, than in rope nurseries where mainly the rope is all that needs regular cleaning of competing organisms. Consequently, routine maintenance to combat fouling (see Maintenance section below) can be one of the largest costs of farming corals *in situ*. At sites with good water quality and where fisheries are managed such that there are plenty of herbivorous fish and invertebrates to control algae, fouling can be fairly easily managed. At sites with lots of suspended organic particulate matter and nutrients and where fisheries are depleted, algae and filter feeding invertebrates can rapidly get out of control and overgrow and kill corals. "Environmentally friendly" anti-fouling paint applied prior to deployment to the main plastic pipe frames of the nursery and the mesh of coral trays can be used to reduce the amount of fouling in the nursery and thus reduce maintenance costs.

We found that a cuprous oxide based anti-fouling paint used in the fish farming industry (Steen-Hansen Maling, Aqua-guard M250) was effective at reducing fouling on and around farmed corals. When the paint was applied to surfaces at least 2 cm away from the coral colonies in the nursery, the anti-fouling agent significantly reduced the amount of fouling of the nursery mesh trays and frames, without harming the farmed colonies[7]. However, when the paint was in contact with coral tissues, it caused bleaching and increased mortality of the corals. The results of the tests revealed that prudent use of limited quantities of the less toxic (more environmentally-friendly) anti-foulants available can be of significant help in reducing maintenance needs and costs.

Use of anti-fouling paint to reduce maintenance and cleaning. Anti-fouling paint has only been applied to the plastic mesh but the plastic pins on which the corals are being grown have been inserted deeper into the mesh to reduce their fouling. Corals are kept at least 2 cm away from the paint (S. Shafir).

Table 4.1 Some pros and cons of fixed and floating nurseries.

Issues	Fixed Nursery	Floating Nursery
Sedimentation	Resuspended sediment can be a problem. If opting for a fixed nursery it is recommended to elevate the coral culture area at least 1 m above the seabed.	There tends to be more water movement when the nursery is suspended in mid-water. This facilitates the washing off of sediment. The height of a floating nursery above the seafloor can be adjusted depending on amount of sediment resuspension.
Light regime	In a fixed nursery all coral species will have the same light regime. The nursery depth will be a compromise that provides enough light for good growth of the corals but minimises susceptibility to bleaching events and waves.	In floating nurseries the depth can be adjusted seasonally to optimize growth conditions (light regime, sedimentation rates) and to avoid excessive irradiation when sea temperatures rise during bleaching events.
Nursery construction	Fixed nurseries have been proved to be more durable in strong currents and their construction and maintenance is simpler.	Floating nurseries are more vulnerable to storms but can be lowered several metres to reduce wave impacts in the event of a storm approaching.
Bottom anchoring	Attachment of fixed nurseries is generally a relatively easy task.	Anchoring of a floating nursery is generally a much more complicated and expensive task than anchoring a fixed nursery to the seabed.
Water flow	Fixed nurseries will tend to be in more protected areas with possibly poor water exchange especially during neap tides.	The continual movement of floating nurseries with the waves is postulated to provide better circulation of nutrients and gas exchange so that coral growth is enhanced.
Proximity to the reef	Fixed nurseries are close to the natural reef and predators (e.g. *Acanthaster*, *Drupella*), disease and resuspended sediment can more easily impact the corals.	Floating nurseries can be set up in deep open water (e.g. at Eilat, where nursery floats at 6–8 m depth above a sandy seabed at 20 m depth). Distance from the natural reef can greatly reduce the negative impacts of natural predators, disease and sedimentation.
Nursery maintenance	Cleaning may need to be carried out slightly more frequently for fixed nurseries and monitoring for predator attacks may also need to be more frequent as they are less isolated. However, ease of access is likely to be better and thus maintenance will be less costly.	Cleaning of floating nurseries may need to be carried out less frequently if water quality is good. Monitoring for coral predators is unlikely to be needed as frequently as for fixed nurseries nearer the reef. Maintenance costs will depend largely on how far you need to travel to the floating nursery and boat requirements.

4

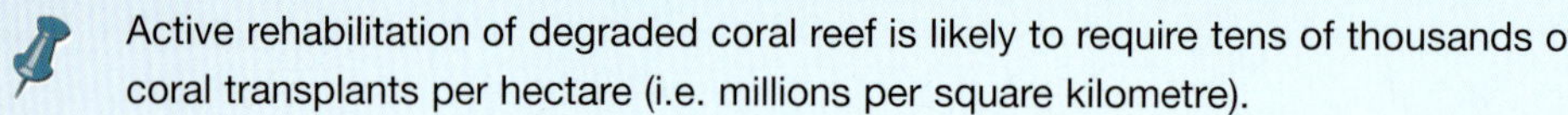

Message Board

- Active rehabilitation of degraded coral reef is likely to require tens of thousands of coral transplants per hectare (i.e. millions per square kilometre).

- Rearing small coral fragments in nurseries allows you to minimise the collateral damage to the natural reef involved in sourcing transplants.

- Simple *in situ* coral nurseries can be constructed from readily available and inexpensive materials and, with some guidance, operated by NGOs or local communities.

- Good site selection is crucial to the success of a coral nursery.

- Above all, conditions (particularly depth) at the nursery site must be appropriate for rearing the types of corals that survive well at the degraded reef which you are planning to rehabilitate.

- Choosing a nursery type (e.g., floating or fixed) and nursery site always involves trade-offs between quality of conditions for rearing, convenience, operating expense and logistics (e.g. available manpower and resources).

- Rope nurseries allow low-cost rearing of large numbers of corals, but as yet methods of effectively deploying corals reared on ropes remain experimental.

- The costs of materials to build a coral nursery are small compared to the costs of stocking and maintaining corals in the nursery and then transplanting them to a degraded reef. It is thus a false economy to try to save money by compromising on quality of materials.

- Rearing corals on substrates that can be fixed quickly and securely to degraded reefs is the key to cost-effective transplantation.

- Unless *in situ* coral nurseries set up to support reef rehabilitation in developing countries can generate income to support their activities (e.g. through the aquarium trade), community-based restoration is unlikely to be sustainable.

- The capability to rear tens or hundreds of clones of selected coral genotypes routinely in coral nurseries provides scientists with a valuable, but largely unexploited, experimental tool.

4.4 Stocking nurseries with coral fragments

The choice of coral species to be reared in a nursery should be dictated largely by the site(s) you wish to rehabilitate. The species must be ones that can thrive at the site to which they are to be transplanted (see Figure 2.3 in Chapter 2). To reduce the risks of mortality from bleaching events and predation and maintain genetic diversity, you should normally rear a mix of species and genotypes (Chapter 3). Fast growing branching species such as acroporids and pocilloporids may generate a rapid increase in structural complexity (and are thus sometimes called "engineering species") but tend to be more vulnerable to both bleaching and predators than slower growing massive, submassive and encrusting species such as poritids and faviids. The environment at the nursery should be sufficiently similar to that at the site being rehabilitated (or more benign) so that transplants will be adequately adapted to the conditions when outplanted.

Collection of material

The nursery may be stocked with natural fragments ("corals of opportunity") or fragments carefully removed from donor colonies. You should ensure that collection does as little collateral damage to healthy reefs as possible.

Corals of opportunity

Corals of opportunity are natural fragments, detached coral colonies, or recruits on unstable substrates (both natural such as coral rubble, and artificial such as ropes and chains) that have little chance of surviving naturally but have a good chance of survival if reared in a nursery, or transplanted directly and securely fixed to the natural reef. Corals fragments resulting from breakage of branching corals are often found lying on the bottom. Massive, submassive, plating and encrusting corals of opportunity may be less common. After storms or destructive human impacts such as ship groundings, corals of opportunity may

Good Practice Checklist

✓ Minimise collateral damage to healthy reefs by using "corals of opportunity" (naturally detached fragments with a low chance of survival) as transplants where feasible.

✓ Always remove dead, moribund or diseased tissue from corals of opportunity before either transplanting directly or rearing in a nursery.

✓ Base your choice of coral species to rear in a nursery on what you expect will survive well at the site you wish to rehabilitate or a comparable "reference" site (section 2.3).

✓ As a precautionary measure, aim to remove no more than 10% of a donor colony if obtaining fragments by pruning coral colonies *in situ*.

✓ Build coral nurseries out of robust and durable materials. Poor construction can lead to the death of thousands of corals.

✓ When deciding on the size of mesh rearing trays, take into consideration how you plan to transport the grown colonies from the nursery to the rehabilitation site. Smaller trays (e.g. 30 cm x 50 cm) are easier to handle and transport.

✓ We recommend using sub-surface rather than surface buoys to support floating nurseries to help reduce interference (e.g. theft of buoys, disturbance of corals by curious divers, etc.) and collision damage by passing boats.

✓ "Environmentally friendly" anti-fouling paint applied to mesh of rearing trays and other structural parts of a nursery can reduce maintenance by about a half.

✓ Rearing substrates for coral fragments should: 1) be cheap and easy to obtain, 2) allow coral fragments to be easily managed and maintained in the nursery, and 3) allow grown colonies to be quickly and easily deployed at the rehabilitation site.

✓ "Cleaning" of the nursery – removal of algae and fouling invertebrates, accumulated sediment, corallivorous organisms, and any diseased corals – must be carried out regularly.

✓ Floating nurseries should be temporarily lowered in the water column when storms or tropical cyclones are forecast.

✓ Use of NOAA sea surface temperature (SST) anomaly charts and inexpensive temperature loggers can allow you to anticipate bleaching events and take measures to try to protect corals being reared in a nursery (e.g. shading, lowering in water column).

be abundant, but it is unclear how abundant or taxonomically diverse they are on the average reef adjacent to areas in need of restoration. A study of five such areas in the Philippines indicated an average of about 1–7 detached fragments per square metre of reef with average geometric mean diameters ranging from 2.4–5.3 cm. About 10 species were represented in a sample of 620 fragments. This snapshot suggests yields of tens of thousands of corals of opportunity per hectare may not be unusual although a rather limited number of common species appeared to dominate.

Corals of opportunity are easily collected and transported to a land or boat facility where they can be prepared for the nursery. Small fragments (c. 2 cm in size) may just need trimming to remove dead, moribund or diseased tissue; larger fragments (over 4–5 cm) may need to be further fragmented (Box 4.2). After trimming, you should gently wash off any debris from the corals of opportunity and fragment further if required. If corals are attached to artificial substrates (mooring ropes, chains, buoys, etc.), they should be carefully detached from them by cutting at the base of attachment with hammer and chisel. If the size of corals of opportunity is above 4–5 cm diameter, corals can be further fragmented as explained below (Box 4.2). If size is below 4–5 cm they can be easily attached (as described below) to substrates for nursery rearing. When attaching small coral fragments to flat substrates for rearing, be sure to smooth the base of the fragment by trimming with an electrician's wire cutter or rubbing it with sand or emery paper. This improves the bonding of the coral to the substrate by the glue. If the corals of opportunity are small recruits of branching species on bits of rubble, we suggest they are left on their substrate and this is bonded to plastic substrate until corals reach 2–3 cm in diameter and can be gently detached at the base and reattached on pins or wall plugs as explained below.

Fragmentation of donor colonies

For branching colonies, carefully cut small branches, 2–5 cm long, from the periphery of the donor colonies using an electrician's wire-cutter or other side-cutting pliers. It is recommended, as a precautionary measure, that you do not prune more than 10% of the donor colony[8]. This reduces the chance of negative impacts on the donor colony. Wear clean cotton (or surgical) gloves and be sure to hold part of the coral colony during the cutting since a strong force applied to the edge of a colony might lead to breakage of a bigger portion than required. Where appropriate try to choose parts of a colony that may be prone to harm in the future (e.g. parts too close to a neighbouring colony or likely to covered by sediment or encroaching macroalgae).

For submassive, massive, encrusting and foliose species up to 10% of the colony (precautionary measure) can be collected using a hammer and chisel. We recommend that you generally take fragments from the edge of colonies to minimise impacts on the donor and facilitate healing. Cuts should be performed diagonally or vertically to the surface of the colony in an effort to minimise the amount of skeleton removed and the amount exposed to attack by boring organisms.

Excision of branches or fragments should be performed underwater with fragments being placed in plastic bags, fine mesh nets, or plastic baskets before being transferred to boat or shore for further processing. You should place each genotype in a separate container to avoid harmful interactions.

Substrates for culturing corals

A variety of substrates have been used for culturing coral fragments in *in-situ* nurseries. These range from slabs of local marble (20 cm x 5 cm x 1.5 cm) to which coral fragments were bound with galvanized steel wire[9], to 10–12 mm diameter plastic wall-plugs into which small (c. 2 cm long) branch fragments are inserted. Over time (often within a few weeks) the coral fragments self-attach to their substrates, growing tissue over the substrate at points of contact, and thus become securely bound to the substrate. In the first case, after several months in the nursery the reared corals on their slabs were wedged into crevices on the reef with their heavy bases providing temporary stability[9]; in the second case after 9–12 months rearing the small coral colonies on their wall-plugs were just inserted into holes drilled into the bare coral rock. Choice of rearing substrate is critical to the cost-effectiveness of transplantation as an intervention.

Inserting a nursery-reared coral colony (c. 8 cm diameter), grown over one year from a 3 cm fragment, into a pre-drilled hole in coral rock at a restoration site.

The following protocols can be used to make coral fragments and nubbins (small fragments approx. 0.25–0.5 cm² in area)[10].

1. The collected coral fragments should be kept wet with clean seawater at all times.

2. For branching species it is recommended that the height of the fragment should be no more than 2–3 times the diameter of the base and should not exceed 3–5 cm. This allows a more stable attachment to rearing substrates and thus lower detachment rates of fragments in the nursery.

3. When fragmenting branching corals of opportunity, you can use branches from all parts of the colony – tip, mid-branch and bases. Additional fragments can be made by cutting the coral pieces using side-cutting pliers, hammer and chisel, or hacksaw.

4. Exposed coral skeleton is susceptible to attack by various organisms. When cutting encrusting/massive species having a thick skeleton with only a thin tissue layer (e.g. *Diploastrea heliopora* or massive *Porites*) it is advisable to reduce the skeleton thickness to allow faster coverage of the exposed area by new tissue and facilitate attachment to the nursery substrate (a hacksaw is effective in trimming excess skeleton).

5. After cutting, wash tissue and skeleton debris from fragments by shaking them in fresh seawater.

6. One way to protect exposed skeleton is to make sure it is coated with epoxy-glue when attaching it to the nursery-rearing substrate.

7. Attach prepared fragments to a nursery-rearing substrate (see below for details on substrates) either by gluing the base with a drop of cyanoacrylate glue ('superglue') to the substrate or by inserting the fragments into plastic wall-plugs or tubing[11].

A faviid fragment that has had exposed skeleton coated with epoxy putty when being attached to a plastic pin for rearing. This helps to protect it from attack by organisms such as boring sponges (S. Shafir).

8. To accelerate attachment of the coral fragment to the substrate, you can sprinkle a thin layer of baking soda (sodium bicarbonate) on the substrate prior to applying a drop of superglue. This can be helpful when attaching larger or problematic fragments.

9. Do the work out of water (i.e. on the deck of a boat or on shore), but make sure that the corals remain moist and in the shade, and use cotton (or surgical) gloves in order to minimize damage to coral tissues.

Ideally the substrate should be 1) cheap, 2) allow easy attachment of fragments for rearing in the nursery, 3) allow easy maintenance (e.g. removal or algae and other fouling organisms) during the rearing phase, and 4) allow easy deployment of reared corals at the site being rehabilitated.

Three cheap plastic substrates have been trialled extensively in recent years with reasonable success (Box 4.3). Firstly plastic pins with a 2 cm diameter head and a 9 cm long pin (waste products of plastic injection moulding), secondly large (10–12 mm diameter) wall-plugs, and thirdly short segments of plastic hose pipe (or any plastic tubing). The latter two substrates are suitable primarily for branching and submassive species (e.g. *Acropora, Montipora, Pocillopora, Stylophora, Heliopora, Hydnophora, Porites rus, P. nigrescens, P. cylindrica*). The plastic pins have also been used with faviids and massive *Porites*. All these substrates can be inserted in plastic mesh trays, which facilitates management and maintenance of the fragments during rearing. Also all these substrates can be inserted into appropriate sized holes drilled into the reef being rehabilitated.

The advantage of wall-plugs and hosepipe is that you can just insert the coral fragments into the open end of a wall-plug or one end of a piece of hosepipe without the need for glue to fix the fragment. Further, having part of the skeleton inserted into the substrate reduces detachment. This speeds up the nursery stocking process and reduces its cost. The disadvantage is that it is only suitable for fairly thin fragments from corals with a branching morphology. The flat-headed plastic pins allow a wider range of morphologies to be attached, but require the use of an adhesive to bond these to the pin head.

Left: *a 9 cm long plastic pin (waste product of injection moulding).* Centre left: *Coral fragments being mounted on pins with superglue prior to immersion.* Centre right: *10 mm wall-plug with a freshly inserted* Millepora dichotoma *fragment.* Right: *A fragment of* Acropora variabilis *being reared on a plastic wall-plug substrate. Note how the basal tissues have grown over the wall-plug. (S. Shafir).*

This can be more time consuming and thus expensive. Also, until self-attachment by growth of coral tissue over the pin head, corals remain more vulnerable to detachment.

Attachment of plating, encrusting and massive fragments directly to plastic mesh with superglue is currently being experimented with and appears promising. Once the fragments have grown in the nursery to a size suitable for outplanting, then the mesh can be cut into patches around each colony and the mesh nailed to the reef using broad headed masonry nails or large galvanised staples. Segments of PVC pipe have also been tested as substrates for massive and encrusting species. Each piece of pipe has holes drilled at diagonally opposed corners; these allow you to tie the substrate to the mesh trays during rearing using cable ties or plastic coated wire, and can be later used to anchor the substrate to the reef using masonry nails or large staples when outplanting at the rehabilitation site.

Several species of massive corals (Favites, Platygyra, Favia) *attached to segments of black PVC pipe with superglue prior to immersion in the nursery. The plastic segments are attached to the mesh of the rearing tray with plastic cable-ties or plastic coasted wire (S. Shafir).*

Arrangement and spacing of corals in the nursery

The sessile life-styles and the growth forms of corals can lead to tissue contacts between adjacent colonies. On the one hand, if these contacts are between fragments from the same donor coral colony (isogeneic ramets) there will be fusions between the colonies, and these will later need to be separated. On the other hand, if these contacts are between different genotypes of the same species (allogeneic conspecifics) or different species (xenogeneic) there can be a striking array of interactions, including nematocyst discharge, development of mesenterial filaments and sweeper tentacles, release of chemicals that inhibit growth, or overgrowth. The net result is that for each interaction there may be deleterious outcomes (e.g., partial mortality, slower growth) for one partner in the interaction. Although some research has been done to look at which species are compatible and which are not, it is easier just to space farmed corals in the nursery to avoid tissue contact.

How long to culture corals in the nursery?

Because of the costs of maintaining corals in a nursery, you generally wish to transplant them as soon as they are large enough to have a good chance of survival on the reef targeted for rehabilitation. This size will vary from site to site depending on ambient conditions (e.g. water quality, herbivory levels) and will also vary with species. Slow growing massive and submassive species appear to survive transplantation better than faster growing branching species and may be transplantable at smaller sizes. Based on the few data available, we would recommend outplanting branching species at around 7–10 cm diameter and massive, sub-massive and encrusting species at around 4–5 cm. At such sizes they will have passed through the stage when they are vulnerable to being destroyed by a single bite from a predator, and transplants of a range of species have shown good survival at these sizes.

For branching coral species, 7–10 cm colonies can be

grown from c. 3 cm high fragments within 9–12 months[12]. At such a size the colonies should be suitable for transplantation to rehabilitation sites. For massive, sub-massive and encrusting species, 4–5 cm colonies can be grown from c. 2 cm diameter fragments in about 12–15 months. Figure 4.7 shows a simulation of various growth rates which can assist you in deciding spacing. The horizontal lines at 5 cm and 10 cm in Figure 4.7 give you an indication of how many months of growth from a 2 cm starting size you could achieve without interference between adjacent colonies for these two initial spacings for the various rates illustrated.

Some spacing will occur naturally as some corals will be lost due to mortality and others may become detached. These losses can be factored in when deciding on initial spacing. Experience shows that barring a mass bleaching or storm event, with good nursery maintenance most mortality and detachment tends to occur in the first two months of rearing. Early attempts at nursery rearing of corals on plastic pins led to detachment losses for *Pocillopora damicornis* and *Stylophora pistillata* of 18%, for *Acropora* spp. of 27% and for *Millepora dichotoma* of 33%. Learning from this to develop better techniques (see Box 4.2 and 4.3) and other substrates such as wall-plugs and hosepipe

segments for branching species has reduced expected detachment to around 5% on average per year. Mortality rates in nurseries with good husbandry appear to average about 10–15% after one year with rates for individual species varying from <5% to 35%. Thus, overall you may expect losses of 15–20% (10–15% mortality plus 5% detachment), which will allow some inherent spacing over time. For example, a 60 cm x 80 cm rearing tray initially stocked with 10 rows of 7 fragments might be expected to lose 10–14 fragments on average.

4.5 Nursery maintenance

For all nurseries, you should check the structure periodically (preferably on a weekly basis) to ensure that ropes have not become frayed, anchors or buoys lost, and that the structure remains robust. In case of any deterioration, materials should be promptly replaced. In floating nurseries, you should regularly check the buoyancy and, if necessary, adjust it. Although use of local natural products (e.g. bamboo) in nursery construction is appealing, these tend to degrade more quickly and increase maintenance costs. At worst, poor quality components can compromise the structural integrity of the nursery putting all the corals in culture at risk. The primary costs in coral rearing and

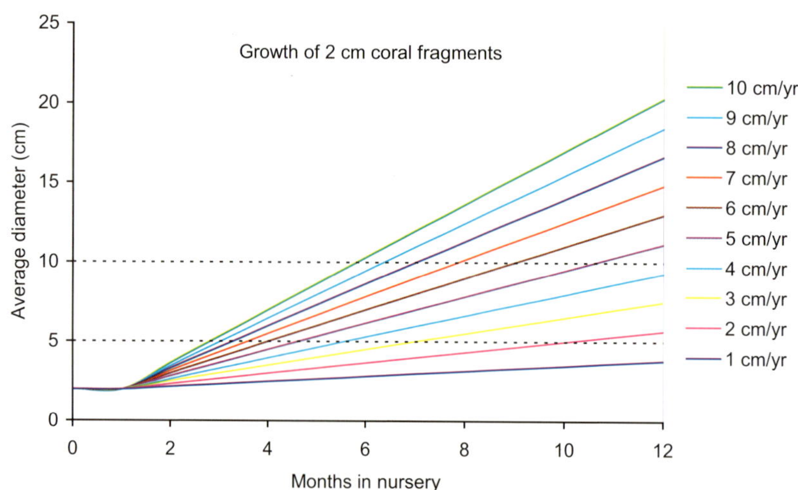

Figure 4.7. A simplistic simulation of growth of 2 cm coral fragments in a nursery for a series of radial extension rates ranging from 1–10 cm per year. Negligible growth is assumed for the first month after transplantation to the nursery and then linear growth in all directions. Slow growing species (radial extension at 1–2 cm/yr) such as the massives *Porites lutea*, the sub-massive *Heliopora coerulea*, or foliaceous *Echinopora lamellosa* could all be spaced initially at about 5 cm apart and not need to be moved apart during a year of rearing. However, faster growing branching species such as pocilloporids follow the 4–6 cm/year trajectory and various *Acropora* spp. increase in diameter in nurseries at 5–6 cm/year and would need to be spaced out half-way through the year if initially only spaced 5 cm apart. However, if initially spaced at 10 cm intervals the graph suggests that fragments of these species would only start to interact towards the end of the year, and with natural wastage due to mortality, it is likely that few such corals would need moving within the nursery. The fast growing acroporid *Montipora digitata* follows the 8–9 cm trajectory so would clearly need to be spaced at least 10 cm apart from the outset and would probably reach a transplantable size within 6–8 months.

transplantation are in people's time and boat and diving expenses, not in materials (see Chapter 7) and savings in construction costs are likely to be false economies which can compromise the whole project.

The objective of nursery rearing is to grow large numbers of small coral fragments into small colonies of a size that can thrive on the reef with as little wastage (death of fragments) as possible. The nursery environment like that in a greenhouse used to rear young plants is predicated to be relatively benign to allow the corals to flourish. Careful choice of the site for the nursery (section 4.2) plays a large role in this but fouling by algae and a range of sessile invertebrates is a significant problem at any *in-situ* nursery, however well-sited. Algae may cause shading and slow coral growth and sessile invertebrates such as sea squirts, sponges, barnacles, and molluscs compete for space and can overgrow and kill corals in the nursery. Just as weeding needs to be carried out in plant nurseries on land, so removal of fouling organisms needs to be carried out in coral nurseries in the sea. One of the largest costs in the two-step coral gardening process is the cost of maintenance to control this fouling. Fouling organisms can both overgrow corals and make the nursery structure heavier and, if the fouling is left unchecked, eventually prone to collapse.

The amount of cleaning of the nursery from fouling organisms that will be needed depends on several factors:

1. *Water quality at the site* – nutrient enrichment can accelerate fouling by algae and high levels of particulate organic matter can favour growth of filter feeding organisms such as sponges, sea squirts, barnacles and bivalves.

2. *Season* – the rate of fouling may be seasonally dependent so that more frequent cleaning is needed at certain times of year.

3. *Predators of fouling organisms present* – the numbers of grazers and invertivores at the nursery and adjacent natural reefs will differ from site to site. These

Diver carrying out routine maintenance of corals being reared on trays in a floating nursery. Note the extensive fouling by algae and other organisms of some trays which have not yet been cleaned (S. Shafir).

organisms have the capability to reduce dramatically the need for human maintenance by consuming potentially harmful fouling organisms which compete with the small corals for space and can overgrow them. Herbivorous fish and invertebrates such as surgeonfishes (Acanthuridae), rabbitfishes (*Siganus* spp.) and the urchin (*Diadema* spp.) reduce levels of algae. Snail predators such as wrasses (e.g., certain *Thalassoma* spp. eat small (<15 mm) *Drupella* whereas *Coris* spp. can eat them up to 25 mm length) may reduce the presence of the coral-eating snail *Drupella* in the nursery. In areas which have experienced severe overfishing, a lack of herbivores and predators of corallivorous invertebrates can exacerbate maintenance problems. Introduction of small urchins (*Diadema*) or topshells (e.g., *Trochus*) to nurseries to assist in maintenance may be helpful.

"Environmentally friendly" anti-fouling paint, applied prior to deployment, can be used to reduce the amount of fouling in the nursery (see section 4.3). However, since it can be harmful to corals too, we recommend that you use it only on parts of the nursery at least 2 cm away from the coral colonies, otherwise corals can suffer increased mortality (25–30%) or detachment[7]. Anti-fouling paint should be thus chosen and applied with extreme care. Use of anti-fouling paint on trays and the nursery structure can reduce maintenance cleaning by almost 50% and thus improve the economics of large-scale restoration.

Cleaning (removal of algae and fouling invertebrates) should be performed more frequently during the early stages of rearing but intervals between cleaning can be increased as corals become larger. Macro-algae are generally best removed by hand (we recommend wearing robust gloves to protect your hands) and encrusting invertebrates may need to be scraped off with a knife. We recommend that cleaning should be carried out once a week for the first 3–4 months, and thereafter on a monthly basis. The time required depends on number of corals and size of the nursery, but we estimate that about 30 hours per month by 2 persons is required to maintain c. 10,000 fragments at a site with good water quality where herbivorous fish and invertebrates are

A rearing tray, which had its mesh painted with anti-fouling paint, after 11 months mariculture in an open-water floating coral nursery. Note the much reduced fouling (S. Shafir).

present. At a site with high sedimentation additional cleaning is likely to be required. Accumulating sediment should always be wafted off using the hands. During spring algal blooms, the amount of maintenance needed may increase considerably. Once every 6 months, you should carry out meticulous cleaning coral by coral, using thin bladed knives and stiff toothbrushes to clean the substrate on which they are growing. You should take care not to damage the corals themselves.

Maintenance should include spacing the growing colonies to avoid fusion between fragments derived from the same genotype (donor colony) or harmful interactions between different genotypes of the same species or coral colonies of different species. Corallivorous snails such as *Drupella* sp. and *Coralliophila* sp. are small and inconspicuous predators and so may be overlooked. These can build up in nurseries until present in sufficient numbers to cause significant damage. The presence of snail predators such as wrasses (e.g. *Thalassoma* spp. and *Coris* spp.), may help to control snail numbers, but in their absence, you should remove the snails from the coral colonies manually using forceps to prevent the build up of damaging infestations. Part of the

Left: *Grazing rabbitfishes (Siganidae) assist in removing algae from a floating nursery.* Right: *Coral tray full of "nubbins" (small fragments) of* Acropora variabilis *about 1-2 months after deployment. Note how the corals have grown basal tissue over the heads of the plastic pins ensuring a good bond with the substrate and are surrounded by wrasses* (Thalassoma rueppellii) *that prey on juvenile corallivorous snails (S. Shafir).*

regular maintenance time of the nursery must be devoted to pest control and removal. We suggest weekly inspection for these snails and other predators with pests being removed from the nursery and eliminated.

Left: *Corallivorous snails (Drupella cornus) destroying a farmed coral colony. The snails on this colony were found too late and the coral died.* Right: *Using forceps to remove a snail (Drupella) which is eating a* Stylophora pistillata *coral colony that has been grown in a floating nursery from a nubbin for 6 months (S. Shafir).*

In addition, you will need to thoroughly clean the farmed corals and the substrates to which they are attached prior to their transplantation onto the reef. About 60 colonies per hour can be prepared by an experienced worker. This task requires considerable experienced labour and a few coral colonies are likely to be damaged during the cleaning process.

Infectious diseases

Infectious diseases and syndromes of corals are caused by microbial agents such as bacteria, fungi, protists and viruses[13]. Poor water quality resulting from anthropogenic inputs (e.g. sewage discharges, aquaculture effluents) and environmental stresses like elevated sea temperatures and high light levels or a combination of these factors may trigger disease outbreaks. In the framework of the increasing need for reef conservation strategies and restoration effort, it is important to identify and deal with emerging coral diseases. However many of the fundamental aspects of these diseases in the wild remain poorly understood. No early warning systems are able to predict outbreaks, and little is known of factors that facilitate disease spread. Lately there has been an increase in the number of coral diseases that have been documented either on the natural reef or in aquarium culture. So far, no outbreaks of coral diseases or syndromes have been positively identified in corals in nursery-culture (although deaths of corals in cage-culture in Palau have been attributed to disease).

Given the possibility of disease outbreaks during restoration work, you should be aware of the symptoms of the most common diseases and simple procedures to avoid spreading disease in nursery facilities. For more detailed information on coral diseases, see the *Coral Disease Handbook. Guidelines for Assessment, Monitoring &*
Management[13] and either the *Underwater Cards for Assessing Coral Health on Caribbean Reefs* or the *Underwater Cards for Assessing Coral Health on Indo-Pacific Reefs* (see www.gefcoral.org for details of how to obtain these) depending on where in the world you are working. These publications provide good photographs of all the common diseases and decision trees to help you to distinguish disease from the effects of (a) tissue loss due to predation by invertebrates and fish, (b) bleaching, (c) invertebrate galls, and (d) other non-disease factors.

If you observe symptoms of disease in maricultured corals, you should remove the diseased fragments from the nursery and, if feasible, keep them in quarantine tanks to monitor disease development. When preparing fragments to stock coral nurseries, you should pay special attention to the health of the donor colonies and not use any fragments which show tissue or skeleton abnormalities in the nursery.

Dealing with storms and bleaching events

Global climate change scenarios forecast an increase in the intensity and frequency of catastrophic events that will damage coral reefs. As natural catastrophes impact on the whole reef, both coral colonies being reared in nurseries and coral transplants on the reef may be severely impacted by unpredictable environmental events (e.g., sea temperature anomalies as in Figure 4.1). There is therefore a need for flexible approaches (adaptive management) and project designs that minimize the risks from such events (see Chapter 3).

Floating nurseries should be designed so that they can be lowered a few metres during bleaching events or heavy storms and deployed at sites with sufficient depth to allow this. For example, each year in Eilat (Red Sea) the floating nursery is lowered two metres in the water column during the stormy season by shortening the anchor cables which attach it to the seabed. To avoid structural damage to the nursery this is done by four SCUBA divers working simultaneously at each corner. In areas prone to tropical cyclones, good shelter is essential. (Lowering a nursery to 20–25 m depth for a short period of time (say a week, during passage of a hurricane or typhoon), instead of being installed at a normal depth of say 3–8 m, will not harm the farmed coral colonies.) Recently in Jamaica a coral nursery was shown to be able to support healthy coral colonies and withstand a major hurricane when temporarily lowered from 6 m to 20 m depth during the passage of the hurricane.

For bleaching events triggered by warm water anomalies, preventative action to protect corals in nurseries needs to be taken <u>before</u> the warming, or a combination of warming and irradiance, stresses the farmed corals and triggers bleaching. Normally sea temperatures are anomalously high (when compared to average temperatures in a given month) for several months prior to a mass-bleaching event, but it is only during the warmest months of the year that this causes problems for the corals. Thus early warning is available and a combination of the internet and cheaper temperature loggers give you the potential to protect your investment in coral "seedlings". Reliable underwater temperature loggers are now relatively inexpensive (around US$ 100 for each logger, but a one-off setup cost of another US$ 200 is likely to be needed for software and downloading hardware). Regional warming anomalies can be followed on the US National Atmospheric and Oceanographic Administration's *Current Operational SST Anomaly Charts* website at: www.osdpd.noaa.gov/PSB/EPS/SST/climo.html. If regional sea surface temperatures appear anomalously warm in the months leading up to the warmest time of year, then a local temperature logger can be downloaded weekly (normally they might be downloaded every 3–6 months) to give warning of developing adverse conditions. If sea temperatures appear likely to rise >1°C above mean monthly maxima (average sea surface temperature during hottest month each year), then floating nurseries can be lowered a few metres to reduce irradiance and shallow fixed nurseries can be shaded using 0.5–1.0 cm diameter PVC mesh which blocks about 10–25% of the light. This should reduce stress on the corals being farmed for the few weeks of the warming. You will need to maintain the mesh used for shading carefully to make sure it does not become covered in fouling organisms and silt.

4.6 Other uses of coral nurseries

Nurseries as sources of coral planulae ("larval dispersion hubs")

In areas with poor coral reproduction, a healthy coral nursery can provide a source of coral recruits for nearby downcurrent reefs. In a floating nursery off Eilat (Red Sea), after just two years in nursery conditions, small branches of *Stylophora pistillata* developed female and male gonads and released viable planulae that settled and metamorphosed at rates equal to 5-year old colonies on adjacent natural reef[14]. With a stock of 10,000 colonies, such a nursery could produce and release over 20 million larvae during the 7-month reproductive season of *S. pistillata* at that site. While the number of planulae developed may fluctuate between brooding vs. broadcast species or between different brooding species, the potential of such a 'larval dispersion hub' for enhancing natural recruitment needs to be studied. During the reproductive season, such a floating nursery could be relocated up-current of reefs targeted for restoration, to enhance larval supply. However, given the very high rates of mortality of larvae and newly settled polyps (see Chapter 5), it is unclear whether this would usefully enhance recruitment of juvenile corals.

Coral nurseries as sanctuaries for endangered species

A coral nursery can be also used as a sanctuary for endangered species. In response to the need to rebuild the surviving populations of an endangered coral species, one could establish an underwater nursery dedicated to propagating such a species. This could farm large numbers of fragments with due attention to maintaining or improving the genetic heterogeneity of the coral species concerned. Examples are recent attempts by national authorities and scientists to develop nurseries for staghorn and elkhorn corals (*Acropora cervicornis* and *A. palmata*) in the Florida Keys. Activities are based on the fact that Caribbean acroporids have undergone about a 95% decline in regional abundance since the 1970s, resulting in their inclusion within the threatened category under the U.S. Endangered Species Act in 2006. It is expected that the nursery-reared corals may be used, when sufficiently grown, as source material for transplantation and will provide an expanding coral stock which can also be used in scientific studies.

Selective propagation of resistant genotypes

There is emerging evidence of quite large intra-specific differences in the survival and growth of different genotypes of the same species[15] and a number of groups in different parts of the world are proposing to rear "resistant" corals. Generally, these are being selected empirically as colonies that have survived a mass-bleaching event, which has killed most of their conspecifics at a particular location. The assumption is that the survivors have innate traits (specific to either the coral host or its zooxanthellar symbionts or a combination of the two) that have contributed to the colonies' survival. Possible local environmental factors or size effects (e.g. small coral recruits <20 mm appear to survive bleaching better than adult colonies) are discounted. Careful research into this intraspecific variation and its underlying mechanisms (in the host coral or its clade(s) of zooxanthellae) are needed. Clearly coral nurseries can play a useful role in propagating genotypes (clones) of interest for experimentation while the sexual rearing techniques described in Chapter 5 would allow selective recombination of resistant ecomorphs.

4.7 Sustainable financing

Funding to set up a nursery and produce one or two years of coral transplants for reef rehabilitation fits well within an average three-year development project. However, reefs tend to recover over periods of decades and nurseries that operate over similar periods are needed. Nursery running costs are such that unless there is a continuing source of outside funding they are likely to be unsustainable. For very short-term projects this may not be an issue; for serious large-scale reef rehabilitation or communities with longer term aspirations to restore their reefs incrementally, this is a crucial issue.

A potential solution is to use part of the *in situ* nursery to produce corals (and perhaps other sessile organisms) for the aquarium trade[16] and use the income generated to support the production of corals for restoration purposes. The requirements of the aquarium trade are specialised and very different to those for restoration, however, with good advice, high value products (e.g. large polyped – *Goniopora*, *Euphyllia* and *Trachyphyllia* – and colourful species) might also be reared and sold to support the nursery operating costs. The feasibility of such an approach requires investigation by a team including an economist and expert in the trade in marine ornamental species. At present, the major impediments to any such approach are national laws in many countries prohibiting the trade in such cultured corals and internationally recognised means of certifying such nursery-reared corals.

References

1. Epstein, N., Bak, R.P.M. and Rinkevich, B. (2003) Applying forest restoration principles to coral reef rehabilitation. *Aquatic Conservation: Marine and Freshwater Ecosystems*, 13, 387-395.

2. Rinkevich, B. (2006) Chapter 16. The coral gardening concept and the use of underwater nurseries: lessons learned from silvics and silviculture. Pp. 291-301 in Precht, W.F. (ed.) *Coral Reef Restoration Handbook*. CRC Press, Boca Raton. ISBN 0-8493-2073-9

3. Bongiorni, L., Shafir, S., Angel, D. and Rinkevich, B. (2003) Survival, growth and reproduction of two hermatypic corals subjected to in situ fish farm nutrient enrichment. *Marine Ecology Progress Series*, 253, 137-144.

4. Shaish, L., Levy, G., Gomez, E. and Rinkevich, B. (2008) Fixed and suspended coral nurseries in the Philippines: Establishing the first step in the "gardening concept" of reef restoration. *Journal of Experimental Marine Biology and Ecology*, 358, 86-97.

5. Shafir, S., van Rijn, J. and Rinkevich, B. (2006) A mid-water coral nursery. *Proceedings of the 10th International Coral Reef Symposium*, 1674-1679.
[Download at: www.reefbase.org/download/download.aspx?type=1&docid=12484]

6. Shafir, S. and Rinkevich, B. (2008) Chapter 9. The underwater silviculture approach for reef restoration: an emergent aquaculture theme. Pp. 279-295 in Schwartz, S.H. (ed.) *Aquaculture Research Trends*. Nova Science Publishers, New York. ISBN 978-1-60456-217-0

7. Shafir, S., Abady, S. and Rinkevich, B. (2009) Improved sustainable maintenance for mid-water coral nursery by the application of an anti-fouling agent. *Journal of Experimental Marine Biology and Ecology*, 368, 124-128.

8. Epstein, N., Bak, R.P.M. and Rinkevich, B. (2001) Strategies for gardening denuded coral reef areas: the applicability of using different types of coral material for reef restoration. *Restoration Ecology*, 9 (4), 432-442.

9. Heeger, T. and Sotto, F. (eds) (2000) *Coral Farming: A Tool for Reef Rehabilitation and Community Ecotourism*. German Ministry of Environment (BMU), German Technical Cooperation and Tropical Ecology program (GTZ-TÖB), Philippines. 94 pp.

10. Shafir, S., van Rijn, J. and Rinkevich, B. (2006) Coral nubbins as source material for coral biological research: a prospectus. *Aquaculture*, 259, 444-448.

11. Shafir, S. and Rinkevich, B. (2008) Chapter 33. Mariculture of coral colonies for the public aquarium sector. Pp. 315-318 in Leewis, R.J. and Janse, M. (eds) *Advances in Coral Husbandry in Public Aquariums*. Public Aquarium Husbandry Series, vol. 2. Burgers' Zoo, Arnhem, The Netherlands.

12. Shafir, S., van Rijn, J. and Rinkevich, B. (2006) Steps in the construction of underwater coral nursery, an essential component in reef restoration acts. *Marine Biology*, 149, 679-687.

13. Raymundo, L.J., Couch, C.S. and Harvell, C.D. (Eds) (2008) *Coral Disease Handbook. Guidelines for Assessment, Monitoring & Management*. Coral Reef Targeted Research and Capacity for Management Program, St Lucia, Australia. 121 pp. ISBN 978-1-921317-01-9

14. Amar, K.O. and Rinkevich, B. (2007) A floating mid-water coral nursery as larval dispersion hub: testing and idea. *Marine Biology*, 151, 713-718.

15. Bowden-Kerby, A. (2008) Restoration of threatened *Acropora cervicornis* corals: intraspecific variation as a factor in mortality, growth, and self-attachment. *Proceedings of the 11th International Coral Reef Symposium*, 1194-1198.
[Download at: www.reefbase.org/download/download.aspx?type=10&docid=13925]

16. Wabnitz, C., Taylor, M., Green, E. and Razak, T. (2003) *From Ocean to Aquarium*. UNEP-WCMC, Cambridge, UK. 64 pp. ISBN 92-807-2363-4
[Download at: www.unep-wcmc.org/resources/publications/UNEP_WCMC_bio_series/17.htm]

Chapter 5.

Rearing coral larvae for reef rehabilitation

James Guest, Andrew Heyward, Makoto Omori,
Kenji Iwao, Aileen Morse and Charlie Boch

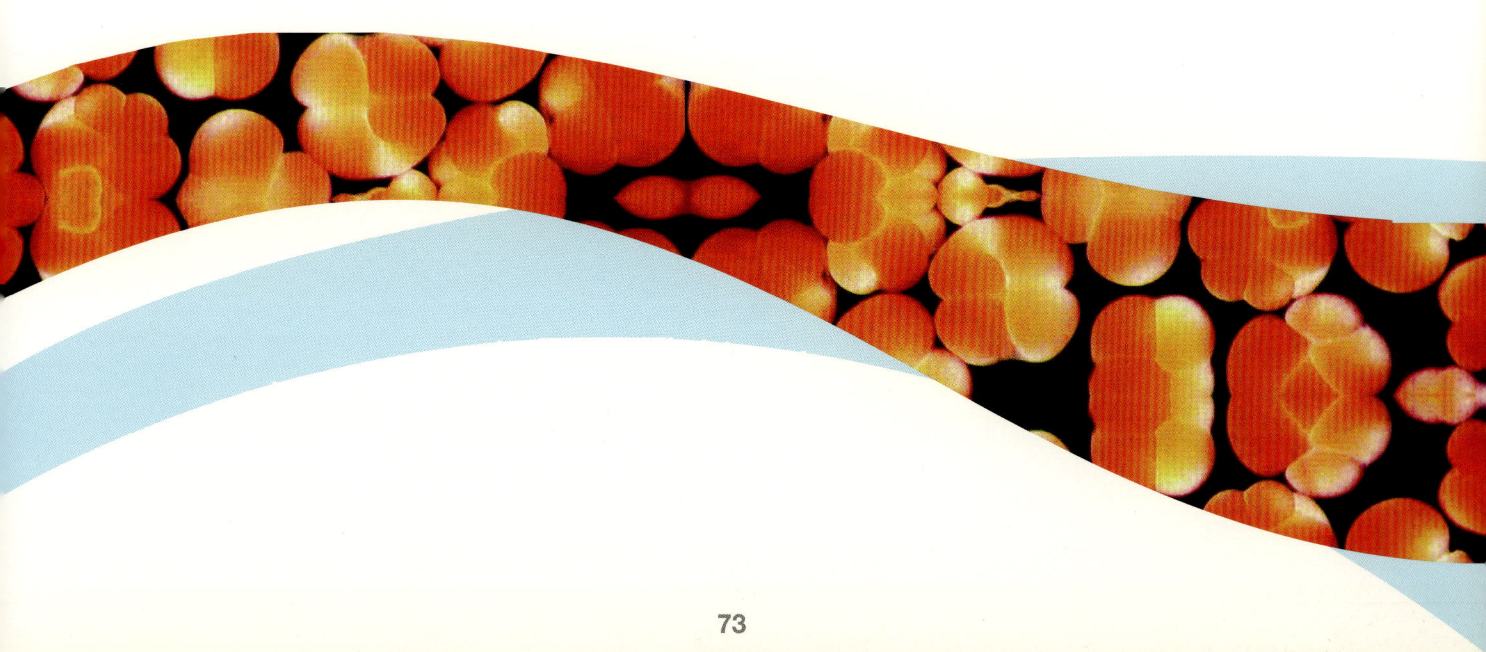

5.1 Introduction

Currently there are few detailed publications offering guidance on how to rear and maintain corals produced by sexual reproduction at a scale suitable for restoration[1-4]. If you are seeking to rear coral larvae *en masse* as part of a rehabilitation project then involvement of a trained coral biologist is strongly advised, especially if this is a first attempt at your location. Although major advances have been made over the last few years, use of sexually reared corals in restoration is still at a largely experimental stage and at the moment appears more costly than the asexual propagation techniques described in Chapter 4.

Furthermore, techniques using asexual propagation have been practised on many species for several decades; by contrast, the various procedures outlined below have only recently been used for restoration and are currently more limited in both the scale of application and the diversity of species being used. Therefore this chapter is aimed primarily at researchers wishing to apply or further develop the techniques for restoration purposes. However, it also aims to provide a summary of the state-of-the-art for restoration practitioners and managers who are considering rehabilitation options using sexually reared corals. If you are not familiar with coral reproductive biology, an overview of the topic is given in Box 5.1.

Contrasting life cycles of broadcast spawning and brooding corals

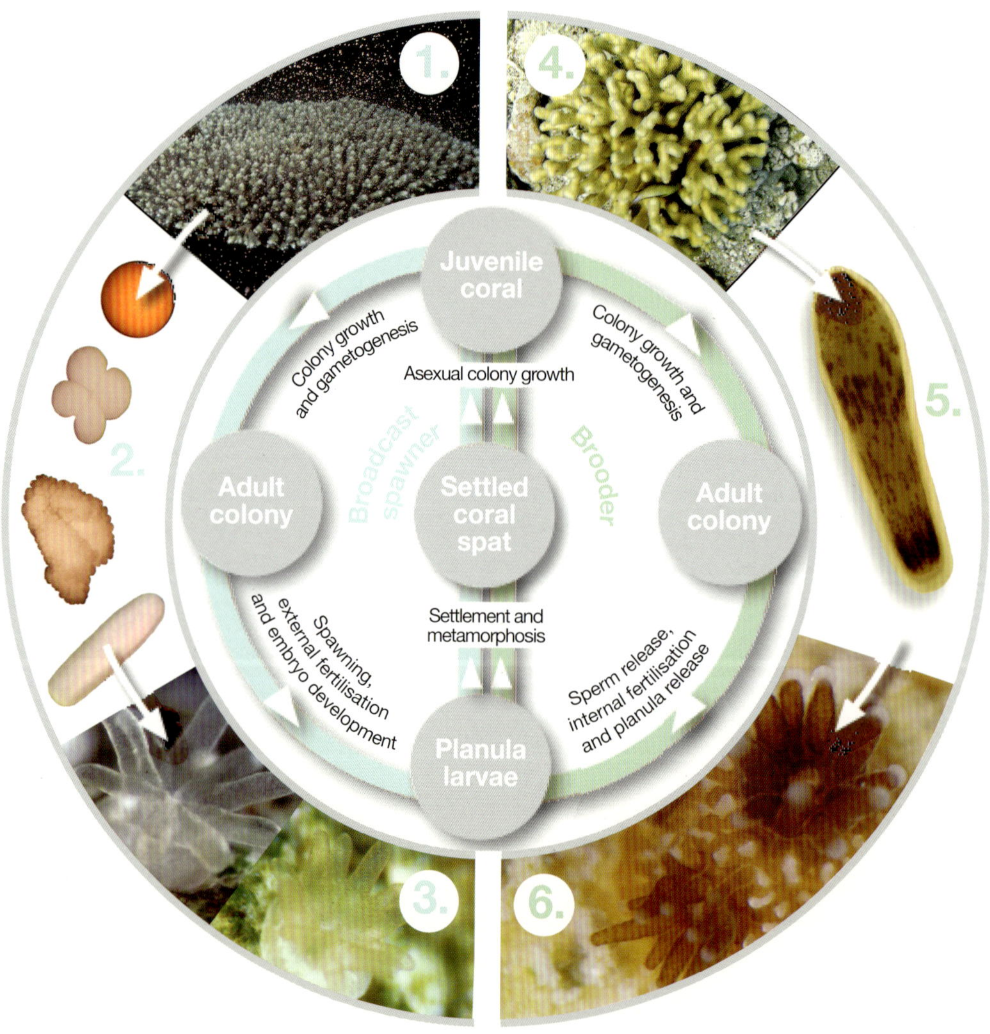

Figure 5.1. Contrasting life cycles of broadcast spawning and brooding corals. **Broadcasters**: 1) mature colonies, such as this *Acropora* colony, broadcast spawn many thousands of buoyant 'bundles' of eggs and sperm at night; 2) fertilisation between eggs and sperm from different colonies and subsequent embryo development occur in the water column (see Figure 5.4 for details); 3) after several days larvae settle on available hard substrata and metamorphose to become azooxanthellate polyps (left photo) and soon after acquire zooxanthellae (right photo). **Brooders**: 4) mature colonies of brooding species, such as *Stylophora pistillata*, spawn sperm that fertilise eggs inside other polyps; these eggs then develop internally and are released as fully formed planular larvae 5) which are usually zooxanthellate; 6) planulae settle and metamorphose within hours or days after release, in the case of some species, such as *Pocillopora damicornis*, settlement can take place very rapidly (i.e. within minutes or hours of release). (Photos 1, 3 and 6: M. Omori and S. Harii; Photo 2: N. Okubo, C. Boch and M. Omori; Photos 4 and 5: B. Linden/B. Rinkevich Lab.)

 An introduction to coral reproductive biology

Corals have two different reproductive strategies that are important to coral larval rearing: broadcast spawning and brooding. Broadcasting species release eggs and sperm (collectively known as gametes) into the water column for external fertilisation and subsequent larval development, whereas for brooders fertilisation occurs within the polyp and fully formed larvae (called planulae) are released during spawning. Broadcasters usually spawn only once each year, while brooding corals may reproduce more than once during a year and often reproduce for several consecutive months. Corals may also be either hermaphrodites (polyps produce both eggs and sperm) or gonochoric (polyps have separate sexes). The majority of corals studied to date are hermaphroditic broadcast spawners (~63% of species) while the remaining species are gonochoric broadcasters (~22%), hermaphroditic brooders (~8%) or gonochoric brooders (~7%) (Table 5.1)[5].

Regardless of the reproductive strategy of the parent, eggs (known as oocytes) and sperm develop inside or attached to filaments deep in the gut of the coral polyps. This process, known as gametogenesis, usually takes several months. So for many corals only one reproductive cycle occurs each year and spawning occurs during a single annual event. The numbers of eggs or offspring an individual can produce is known as its fecundity. Large colonies may have thousands of polyps and each polyp may contain many oocytes, so corals have the potential to be highly fecund. There are considerable differences among species in the age and size of colonies at sexual maturity, but many corals become reproductively mature within 1 to 5 years after settlement. In some cases very small colonies can be reproductive (e.g. 3 cm diameter colonies of *Stylophora pistillata* and 4 cm diameter colonies of *Favia favus*), however, sexual maturity is only found consistently in larger colonies ranging from 6 cm (e.g. *S. pistillata*) to 15 cm diameter (e.g. *Acropora hyacinthus*)[6].

Life cycles of broadcasting and brooding corals

Broadcasters

Whether hermaphroditic or gonochoric, broadcasting corals release gametes into the water column where they will meet gametes of other colonies. For hermaphroditic species, eggs and sperm are usually bundled together as buoyant packages at the time of release from the polyp. The buoyancy of the bundles brings them to the surface where they break apart, releasing eggs and sperm at the sea surface and allowing cross fertilisation to occur. If fertilisation is successful, cell division can be observed within one to two hours and fully developed larvae form as early as two days after fertilisation. Planulae are then able to settle and metamorphose into polyps (Figure 5.1). Newly settled polyps begin to deposit a calcium carbonate skeleton and acquire zooxanthellae (although for some species zooxanthellae are transferred to the eggs from the parents prior to spawning, or planulae take up the symbionts prior to settlement). Asexual production of new polyps and calcification continue so that polyps form adult reproductively mature colonies within a few years. Broadcast spawning corals often release gametes synchronously, that is, members of the same species spawn at the same time to achieve cross fertilisation. In many cases, corals of two or more species spawn in very large numbers during 'mass synchronous spawning events'. During mass spawning or shortly after, it is sometimes possible to find 'slicks' containing many millions off eggs, sperm and embryos floating on the sea surface. Indeed spawn slicks can be used as a source of embryos or planula larvae for rearing.

Brooders

In contrast to broadcasters, brooders take up sperm released from nearby colonies and following internal fertilisation of the eggs subsequently release fully formed planulae. Planulae are released by polyps directly into the water or occasionally they are externally brooded in a specialised pouch on the surface of the coral. Brooded larvae usually contain zooxanthellae, are often able to settle and metamorphose within minutes of release and can begin feeding immediately (Figure 5.1). Some brooding species can produce planulae that are apparently asexually formed. In this case, genetic diversity is not enhanced, but a specialised genotype is produced that is likely to be well adapted to local conditions.

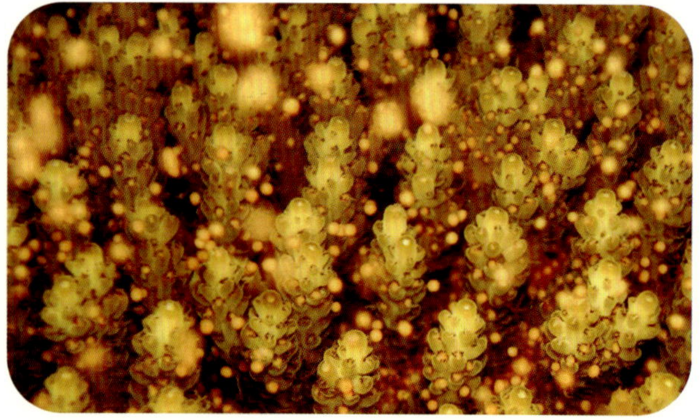

A hermaphroditic branching coral (Acropora sp.) releasing bundles of eggs and sperm (known as gamete bundles) at night in the Philippines (K. Vicentuan).

A hermaphroditic massive coral (Montastrea colemani) releasing gamete bundles in the Philippines (J. Guest).

A female gonochoric coral (Euphyllia ancora) releasing eggs at night in Singapore (J Guest).

A male gonochoric coral (Goniopora sp.) releasing clouds of sperm at night in the Philippines (J. Guest).

Table 5.1 Summary of the dominant reproductive mode for common Indo-Pacific and Atlantic scleractinian coral families. This table is a simplified guide and each family contains more than one reproductive mode[5]. Families are from Veron (2000)[7] however scleractinian phylogeny is being revised based on molecular evidence and many of these families are now considered to be split into several phylogenetic clades[8].

Hermaphroditic broadcasters	Gonochoric broadcasters	Brooders
Acroporidae[a]	Agariciidae (Indo-Pacific)	Agariciidae (Atlantic)
Faviidae[b]	Astrocoeniidae[e]	Astrocoeniidae[e]
Merulinidae	Dendrophylliidae[f]	Dendrophylliidae[f]
Mussidae (Indo-Pacific)[c]	Euphylliidae[g]	Meandrinidae
Oculinidae[d]	Fungiidae[h]	Mussidae (Atlantic)[c]
Pectiniidae	Poritidae[i] (Indo-Pacific)	Siderastreidae[j]
	Siderastreidae[j]	Pocilloporidae[k]
		Poritidae[i] (Atlantic)

Exceptions: a) Acroporidae: *Isopora bruggemanni, I. cuneata, I. palifera* and *I. togianensis* (hermaphroditic brooders); b) Faviidae: *Cyphastrea ocellina* (hermaphroditic brooder), *Goniastrea aspera* (can brood and broadcast), *Diploastrea heliopora* (gonochoric broadcaster and is a now considered to be in a different phylogenetic clade)[8], genus *Leptastrea* (contains gonochoric broadcasters and is now considered to be grouped with family Fungiidae)[8]; c) Mussidae: this family is now considered to consist of several different phylogenetic clades, genus *Mussimilia* (endemic to the South Atlantic) comprises hermaphroditic broadcasters; d) Oculinidae: *Galaxea fascicularis* and *G. astreata* (pseudo-gynodioecious, i.e. female colonies release eggs and male colonies release sperm packaged with non-viable eggs, *Galaxea archelia* (hermaphroditic brooder), genus *Galaxea* (now considered to be grouped with family Euphylliidae)[8]; e) Astrocoeniidae: genus *Stephanocoenia* (gonochoric broadcasters), genera *Madracis* and *Stylocoeniella* (hermaphroditic brooders and are now grouped with Pocilloporidae)[8]; f) Dendrophylliidae: *Dendrophyllia, Tubastrea, Balanophyllia* (brooders), *Heteropsammia* and *Turbinaria* (gonochoric broadcasters); g) Euphyllidae: *Euphyllia glabrescens* (hermaphroditic brooder; h) Fungiidae: *Heliofungia actiniformis* (capable of both brooding and broadcasting)[5]; i) Poritidae: genus *Porites* (contains several gonochoric and hermaphroditic brooding species), genus *Alveopora* (contains both hermaphroditic brooders and broadcasters and is now considered to be grouped with family Acroporidae)[8]; j) Siderastreidae: *Siderastrea siderea* (broadcast spawner), *S. radians* and *S. stellata* (gonochoric brooders), *Coscinaraea columna* and *Psammacora stellata* (gonochoric broadcasters and these genera are now grouped with family Fungiidae)[8]; k) Pocilloporidae: *Pocillopora damicornis* (capable of brooding and broadcasting), *P. elegans, P. eydouxi, P. meandrina* and *P. verrucosa* (hermaphroditic broadcasters).

5.2 Importance of differences among species and locations

Most research on larval rearing for reef restoration has focused on relatively few species. Research on broadcasters has concentrated on the genus *Acropora* and a few members of the family Faviidae, while that on brooding corals has been done primarily with the species *Pocillopora damicornis*, *Stylophora pistillata* and *Agaricia humilis*. There are several important differences among species that influence the larval rearing techniques used; these include egg size and buoyancy, embryo developmental rates, time until larvae become competent to settle and whether eggs are zooxanthellate or non-zooxanthellate (most are non-zooxanthellate). Temperature also affects developmental rates and therefore these will vary among locations. Nonetheless, for more than two decades coral larvae from many species and families have been reared successfully for experimental purposes using essentially the same techniques. Consequently, the methods outlined below provide a sound basis for carrying out larval rearing for restoration for a range of species.

In using sexually reared corals for restoration, different approaches are likely to be needed on Atlantic versus Indo-Pacific reefs. In the Atlantic, almost half of the species studied are brooders compared to less than one fifth of Indo-Pacific species. In the Indo-West Pacific, hermaphroditic broadcast spawning species are more common (75% of species) compared to Eastern Pacific reefs which have a higher proportion of gonochoric broadcasters (>70% of species)[5]. Difficulties may arise when attempting to rear larvae of gonochoric species in the laboratory (e.g. families Agariciidae, Astrocoeniidae, Dendrophylliidae, Euphylliidae, Fungiidae, Poritidae and Siderastreidae) (Table 5.1) as a mix of males and females is required and sex in corals cannot be determined easily in the field.

5.3 Rationale for using larvae for reef restoration

There are several reasons for using reared larvae for reef rehabilitation. Firstly, restoration based on sexual propagation will tend to result in higher genetic diversity than that from asexual propagation (Chapter 4). Secondly, corals are highly fecund, thus coral larval rearing has the potential to produce very large numbers of juvenile corals if the normally high levels of early mortality are reduced. Finally, in an ideal situation, larval rearing should result in little damage to existing reefs as "donor" colonies can be returned to the wild after spawning. Larval rearing however is much more labour intensive compared to asexual techniques, requires additional expertise, facilities and accurate information on spawning seasonality and timing.

Corals may broadcast gametes for external fertilisation or internally brood larvae (Box 5.1), and both strategies can be exploited to generate larvae for restoration purposes. Both broadcasters and brooders can have very high fecundity, although broadcasters tend to concentrate their annual reproductive output into brief annual spawning periods, while brooders produce fewer larvae but over extended reproductive seasons lasting several months. In nature the vast majority of sexual propagules do not survive with the heaviest mortality occurring during the first few weeks of life. Similarly, during a natural spawning event, fertilisation levels may be high but are often variable and many embryos and developed larvae will fail to reach the stage at which they are ready to settle out of the plankton (competency), attach to the substrate and metamorphose into polyps (settlement and recruitment).

By contrast, over 90% fertilisation success can be achieved consistently in culture and if embryos and larvae are well cared for, the majority of spawned eggs can survive to become fully formed planulae. For broadcast spawners it is normal to obtain tens to hundreds of thousands of eggs from individual sexually mature colonies of 10 to 30 cm in diameter. Consequently, with several broadcasting colonies, it is possible to generate more than one million larvae and even greater numbers can be harvested from spawn slicks. Brooders on the other hand can release several hundred to several thousand planulae during a spawning season but the larvae are fully formed upon release so are comparatively robust.

If done with care, larval rearing methods should result in little damage to the existing reef, particularly when spawn is collected directly from the sea, either from collectors placed over spawning colonies or from surface spawn slicks (section 5.5, methods 1 and 2). If spawn slicks are utilised, then natural levels of genetic variation can be attained, whereas genetic mixing will be limited by the number of colonies used when collecting from individuals. Spawn slicks also provide material that is representative of the local coral assemblage with a potentially wide range of species present, however slicks are not always easy to find. Collecting and maintaining broodstock colonies at a land based hatchery for spawning provides an alternative and often more convenient strategy for harvesting gametes, but this requires a good water circulation system and careful handling to minimise mortality before colonies are returned to the wild (section 5.5, method 3).

Reared coral larvae can be used for restoration in two main ways: 1) fully formed coral larvae may be settled onto artificial or natural substrata and reared in aquaria or *in situ* nurseries until they are ready to be out-planted to areas of degraded reef, or 2) larvae can be introduced directly to degraded areas of reef at very high densities by containing the larvae and allowing them to settle naturally.

An alternative approach, which does not require larval rearing, is to allow corals to settle naturally on specially designed coral settlement substrata deployed in areas that receive high recruit densities during coral mass spawning.

After a few months the settlement substrata with naturally settled corals are transplanted to areas of degraded reef[9]. This method is considerably simpler than the first two methods as it does not require coral larvae to be cultured. The focus of this chapter is primarily on larval rearing, therefore this method will not be covered here but further details are given in case study 7 (Chapter 8). All of the outlined methods are being tested at the moment and some promising results are emerging, however considerable research is still needed before sexual propagation methods are used in applied restoration efforts. It is hoped that this chapter will stimulate further research and development of these techniques.

5.4 Identifying when corals are ready to reproduce

Before you can attempt to rear coral larvae, you need to obtain accurate information on the timing of coral reproduction for your location. Information exists about the seasonal timing of coral spawning and planular release for most regions (Figure 5.2), however detailed information on the exact timing of reproduction are lacking for many areas. You can consult the scientific literature, web-based discussion lists (e.g. Coral-List: http://coral.aoml.noaa.gov/mailman/listinfo/coral-list/) or even local dive shops and fishers to find information on when local corals have been observed to spawn or release planulae. Fortunately at a given location there is reasonable consistency from one year to the next in terms of the general timing of reproduction at coral community, species and colony levels, so even anecdotal observations can help you in pinpointing exact spawning times.

For the much of the Indo-West Pacific and Western Atlantic, broadcast spawning peaks over two or more consecutive months typically during either spring or autumn. Populations of a single species at a location may either split their spawning over consecutive months or spawn predominantly in one month. Broadcast spawning can be highly synchronised within species with the majority of the population releasing gametes over a few consecutive nights.

Brooding species on the other hand often release planulae (a process called "planulation") during several consecutive months, although some brooders have just one annual cycle (e.g. *Heliopora coerulea*). On high latitude reefs planulation may be confined to just a few months, whereas on low latitude reefs (those between 10°N and 10°S), it may occur throughout the year. Nonetheless, even at low latitudes there tend to be seasonal peaks (i.e. planulation will be greater in some months than others) and these may occur around the time of the seasonal peaks for broadcasting species during the warmer sea temperature months.

The key point is that each location and species may have different spawning patterns; therefore it is essential to have reliable information on the timing of spawning for your location (Box 5.2) in order to source gametes for larval rearing efforts.

Heliopora coerulea, an octocoral that forms sub-massive or plate-like colonies and broods planula larvae on the colony surface (J. Guest).

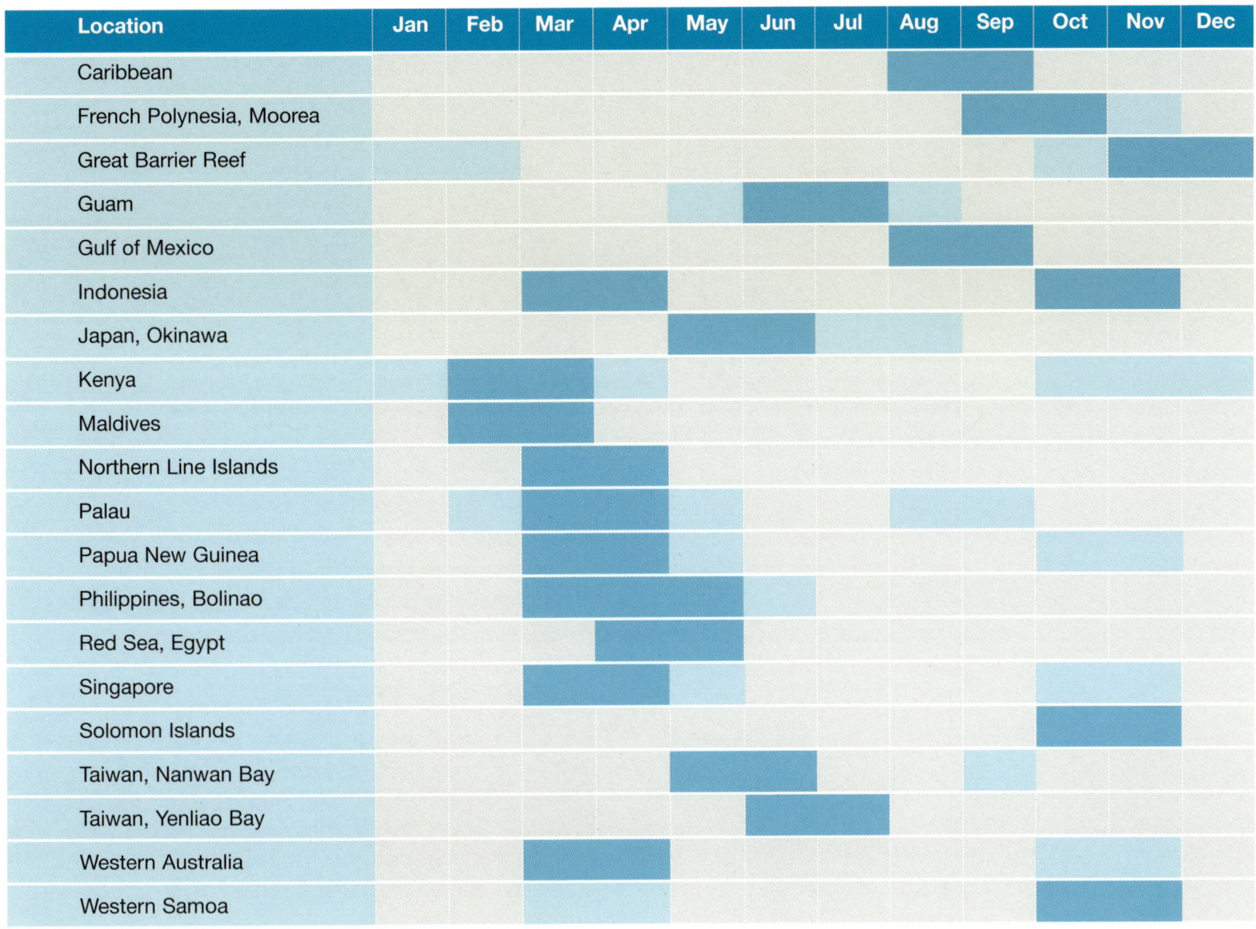

Figure 5.2. The two peak broadcast spawning months for coral reef areas where multi-species spawning has been observed or inferred from existing published studies. Darker blue bars show two peak months, lighter blue bars show minor spawning months.

Box 5.2 What controls the timing of coral reproduction?

The timing of coral spawning and planulation is regulated by several factors and timing varies considerably among species and locations. Reproductive timing is influenced by environmental factors which, in the case of broadcast spawners, act as 'cues' to allow corals to synchronise reproduction so that cross fertilisation can occur. For broadcast spawners, environmental cues are thought to work at progressively finer time scales to select for the time of year, the lunar date and the exact time of spawning in relation to sunset. Sea temperature and sunlight (longer daylight hours and more intense solar radiation) are thought to be involved in regulating the time of year and for many locations the greatest number of corals broadcast spawn just before or during the warmest and calmest months.

Lunar cycles are almost certainly involved in controlling the dates of spawning and planulation for many broadcasting and brooding species. Corals may spawn or planulate at any time during the lunar cycle but the lunar periodicity for a given species and location is often consistent from year to year. In the case of broadcasters, spawning peaks during the week following the full moon in many locations. The exact time of spawning is controlled by the day/night cycle with the majority of spawning activity happening between sunset and midnight for broadcasters, whereas brooders may release planulae at any time during the day or night, although some species have distinct planulation peaks during certain hours.

How to assess proximity to spawning

For brooding species there is no simple way of assessing reproductive timing, however many brooding species reproduce monthly, making it possible to obtain larvae by maintaining corals in aquaria or by deploying collecting devices over colonies *in situ* (see Section 5.7) to monitor timing of planulation.

For broadcasting species there are several methods for assessing proximity to spawning. The simplest and quickest method is to examine egg pigmentation in artificially fractured polyps *in situ*. Eggs of many broadcasting species become pigmented close to the time of spawning. Colours range from red, orange and pink to blue, green and in some cases brown (the brown pigmentation is due to zooxanthellae in eggs of certain species, e.g. *Montipora* and *Porites*). Pigmentation may happen several days or weeks before spawning and this will vary among species and locations. However a general rule is that corals with deeply pigmented eggs will spawn following the next full moon. This certainly appears to be true for many *Acropora* species.

Up to three branches should be removed carefully from a colony with side-cutting pliers or a thin chisel used as a 'lever' to snap off the branch. Branches should be snapped approximately half way between the branch base and tip as this is often where most eggs are found. The tips of branches and the edges of colonies are often sterile zones and will not provide a reliable indicator of reproductive maturity. We have found for *Acropora* species that it is

easier to see mature eggs if you gently lift the fractured branch a few millimetres from the broken base so that strands of mature oocytes are stretched across the gap between the two exposed pieces of skeleton (see photos below). For massive species a similar approach can be used but will require removing pieces from colonies with a hammer and chisel. Due to the overlap of spawning between many species from the dominant coral families, sampling of the easiest to observe colonies (e.g., branching *Acropora*), can often suffice to predict the onset of spawning for other common species.

For species with mature oocytes larger than 300 μm in diameter (e.g. *Acropora*, *Montipora*, and many faviids) it should be possible for divers to observe egg status underwater with the naked eye. For species with small oocytes less than 300 μm in diameter (e.g. Fungiidae, *Porites* spp.) or oocytes that are immature, it may be necessary to inspect freshly collected samples under a dissecting microscope at x32 to x64 magnification.

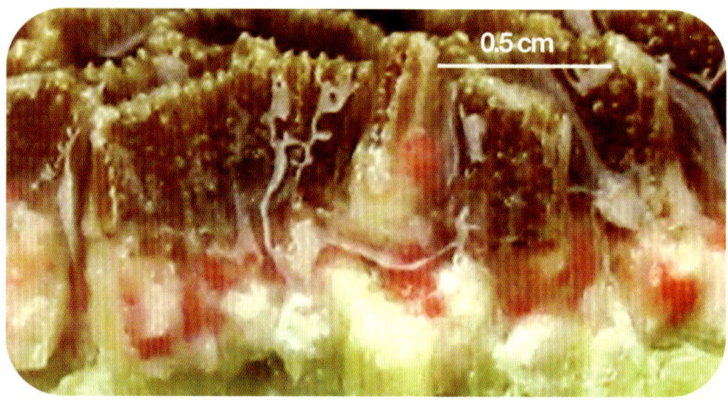

A fragment taken from a massive coral colony (Platygyra sp.) containing red pigmented mature eggs (J. Guest).

A fragmented branch of an Acropora coral colony showing visible white eggs; this indicates that the colony is probably going to spawn within 1 to 3 months (K. Vicentuan).

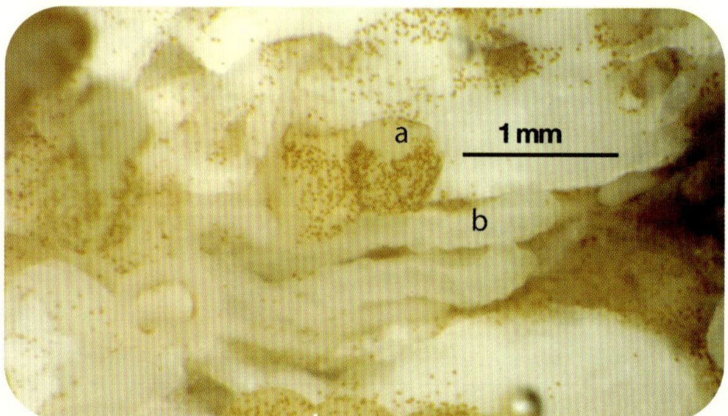

A microscope photograph of a freshly dissected fragment of Montipora sp. showing a) mature eggs (containing zooxanthellae) and b) testes (A. Heyward).

A fragmented branch of an Acropora colony showing deeply pigmented orange mature eggs; this indicates that the colony is likely to spawn soon after the next full moon (K. Vicentuan).

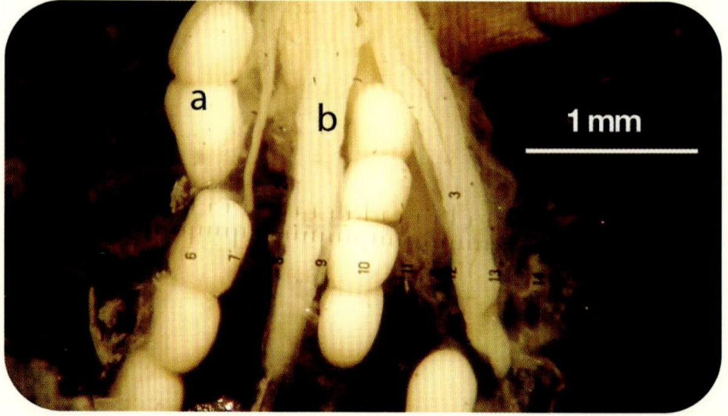

A polyp of Acropora muricata that has been preserved in formalin then decalcified and dissected to reveal a) large mature eggs and b) mature testes (K. Vicentuan).

Other methods for assessing spawning timing are more labour intensive but may be necessary to establish local spawning patterns, particularly when oocytes are smaller than 300 µm. Samples of coral containing several polyps are fixed in 10% seawater formalin for 24 hours then decalcified in a dilute acid (e.g. 10% HCl) until all skeleton has dissolved (this can take several days and may require several changes of the acid solution) and stored in 70% ethanol. Samples should be collected monthly until oocytes become visible in dissections. At this time, samples can be collected more frequently until oocytes disappear from samples. Disappearance of oocytes between sampling days is a strong indication that spawning has occurred.

If colonies appear ready to spawn they can be kept in tanks on land and observed each night for signs of spawning. Water flow should be shut off each night just before sunset and the tanks should be kept in an area shielded from artificial light. Monitoring of colonies approximately every 30 minutes from sunset until midnight should allow detection of most spawning activity, although a few species can spawn in the early hours of the morning. Care must be taken to shield the colonies from too much artificial light during checking. A quick check with a low-intensity flashlight should not disrupt normal spawning and use of a red cellophane filter over the light source will further reduce the likelihood of disrupting spawning behaviour. For some hermaphroditic broadcasting species, gamete bundles appear in the mouths of polyps 1 to 3 hours before release on the night of spawning (this is known as 'bundle setting'). It is worth noting that leaving corals in tanks for extended periods may alter their normal spawning times and may reduce colony health, so if the option is available colonies should be kept in the sea in between observations and only maintained in aquaria as long as is necessary for observation. If reefs are easily accessible then it may be possible to dive at night to observe spawning in the field. Direct observation of spawning is by far the best indicator of normal spawning times; however it may be difficult or impossible to safely carry out night diving in some locations. A simple alternative is to firmly place the base of an upturned clear plastic bottle over colonies a few days before the predicted spawning night and check for the presence of trapped gametes each morning.

Example of a simple gamete trap, consisting of an upturned clear plastic bottle base affixed to a recently spawned colony of Acropora humilis, *used to monitor coral spawning timing in Palau (C. Boch).*

Table 5.2 Relative advantages and disadvantages of using different methods for harvesting coral gametes for larval rearing.

Method	Advantages	Disadvantages
1. Collecting from spawn slicks	Does not require that colonies be removed from the sea so there is no collateral damage. Slicks may contain coral gametes from several species and many colonies such that the level of genetic diversity will be equivalent to that found in nature. Does not require scuba diving or maintaining corals in aquaria which could reduce costs.	Spawn slicks are not a reliable or predictable source of gametes and embryos and slicks do not form in all locations. Often requires boat access to offshore localities. Slicks often contain unwanted debris that may reduce water quality during larval rearing. The range of species present in a spawn slick may have different competency periods and hence rearing and settlement requirements.
2. Collecting from coral colonies *in situ*	Does not require that colonies be removed from the sea so there is no collateral damage. Does not require maintaining corals in aquaria which could reduce costs.	Likely to require scuba diving at night which may be difficult and expensive in certain locations and is prone to disruption if weather conditions are bad. This may be only way in areas where collection of corals from the reef is prohibited. Cannot be used for collecting sperm from male gonochoric colonies.
3. Collecting from colonies *ex situ*	Allows for much greater control of the spawning process and does not require scuba diving which may reduce costs. Is not subject to changes in weather conditions.	Corals may die if they are not well looked after in aquaria or not carefully replaced to the reef. This is labour intensive if colonies need to be reattached to the reef or maintained as brood stock after spawning. Corals must be maintained in flow-through aquaria until spawning which may require significant investment in aquaria facilities.

5.5 Harvesting coral spawn

To rear coral larvae you will need to harvest coral gametes or newly released larvae. There are three main ways to do this (Table 5.2):

1. You can collect gametes and recently fertilised embryos from near the sea surface where they form spawn slicks, or

2. You can collect gametes *in situ* by placing collecting devices over mature colonies, or

3. You can remove mature colonies from the reef and maintain them in aquarium tanks on shore or on board a boat to allow spawning to take place *ex situ*.

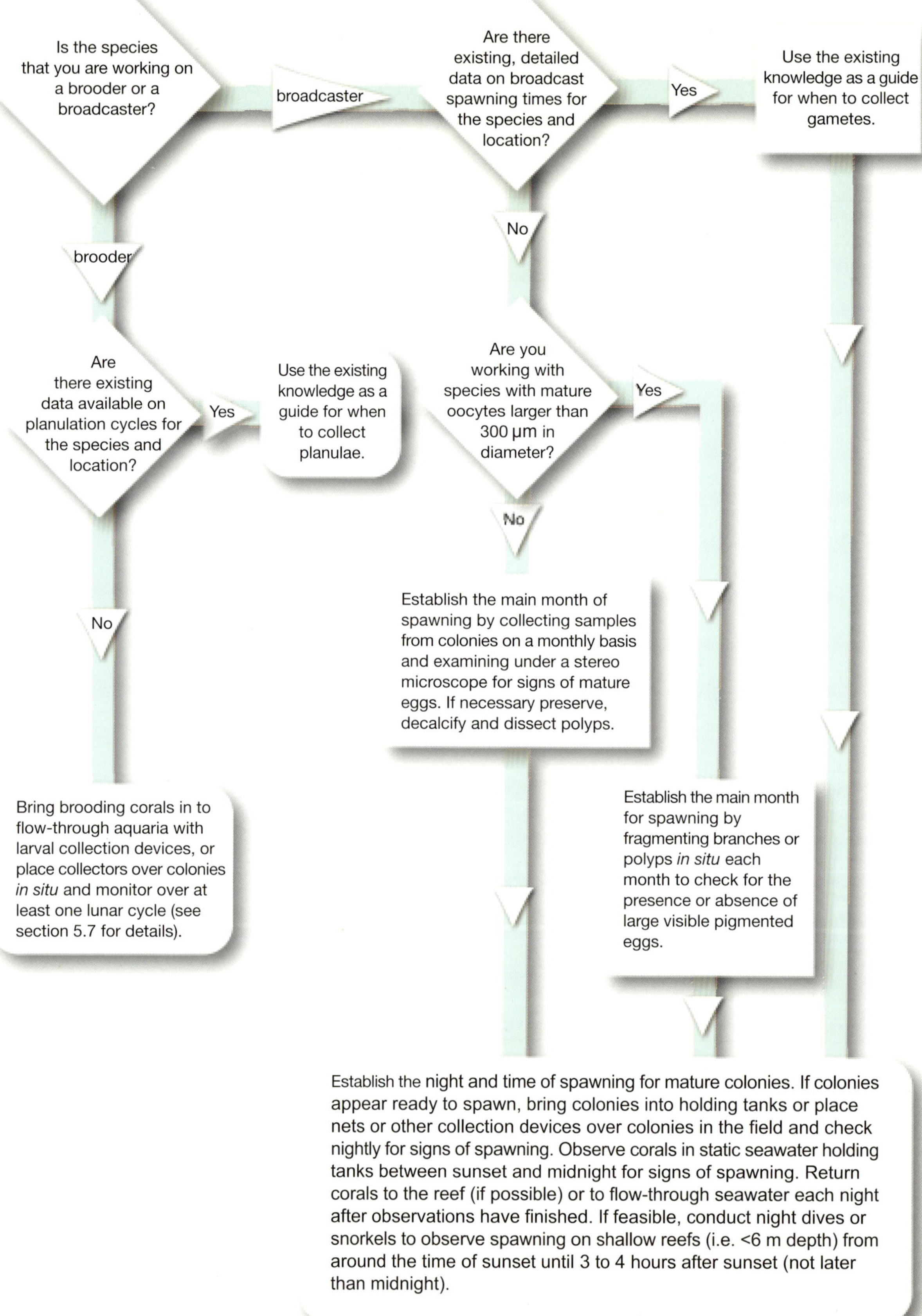

Figure 5.3. Decision tree: how to assess the timing of spawning in broadcasting and brooding corals.

1. Collecting from spawn slicks

Immediately following large mass spawning events, slicks of fertilising eggs and embryos may form at the surface of the sea. Ideally you should collect embryos soon after spawn slicks have formed, however it is possible to collect embryos from offshore slicks the morning after spawning has occurred. More than two hours post-spawning, spawn slicks may begin to disperse; after five hours embryos at the centre of slicks may be non-viable; and more than 17 hours post-spawning slicks may become putrid and contain only non-viable embryos.

Collecting from slicks offshore will reduce the chance of land based pollution, although collected seawater may still contain much debris, dead material and associated bacteria which can reduce the culture water quality and will require extra work to remove. Slicks that have washed up on shore often do not contain living embryos, however it may be possible to collect from slicks in artificial enclosed bays (e.g. from harbour walls) although collection should be done within 2 hours of spawning as many of the embryos will become damaged when they collide with the harbour wall.

You should collect gametes from the edge of the slick by scooping at or just below the sea surface (<0.5 m depth) using plastic buckets or dippers (approx. 2–10 litre volume). The best technique is to slowly depress the bucket or dipper into the sea and allow water containing high numbers of eggs and embryos to gently flow over the bucket lip. When you have collected sufficient numbers of embryos, they should be 'cleaned' by doing water exchanges and transported carefully but quickly back to the laboratory to be counted and stocked in larval rearing tanks (section 5.6).

A spawn slick on the water surface the day after a mass spawning event in Akajima, Japan (K. Iwao).

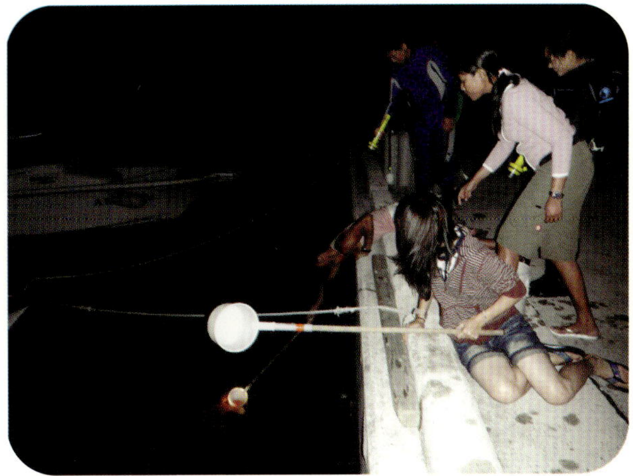

Collecting from a spawn slick at night from a harbour wall using collectors on long arms in Akajima, Japan (M. Hatta).

2. Collecting from coral colonies *in situ*

Using the methods described above in section 5.4, you should check coral colonies to see if they are mature. Only if they contain obvious deeply pigmented eggs should you attempt to collect gametes (using method 2 described here or method 3 described below). If you decide to collect gametes *in situ* then you will need to place collecting devices over several mature colonies just prior to spawning. One approach is to wait until you observe 'bundle setting' in colonies before you attach the nets. This method saves repeated placement of collection nets over corals on successive evenings until spawning happens. Typically, collecting devices consist of a funnel made of plankton mesh (100 µm mesh size for most species) with a transparent plastic bottle (0.5 to 1 litre volume) attached at the mouth of the net. The bottle can be kept afloat and upright by introducing a small amount of air to the upturned bottle or by attaching a piece of polystyrene foam around the bottle's neck.

For broadcast spawners, gamete bundles are usually buoyant and so will float towards the surface and become trapped in the bottle. The net can be made rigid by addition of stainless steel wire supports and this may help prevent the net brushing against the coral due to waves and currents. The collecting device can be attached to the colony or surrounding substrata using various methods including cable ties, stainless steel nails, aluminium caribiners or flexible coated wire. For small colonies a draw string can be used to tighten the base of the net around the colony so that the net encloses the whole colony, for larger colonies the net need only cover a portion of the colony surface. A net of 30 cm diameter should capture several hundred thousand eggs from a mature *Acropora* colony.

When spawning is finished, you should collect and close the bottles underwater and transport the gamete bundles <u>immediately</u> to the fertilisation tanks to be mixed with gametes from other colonies. If gametes are left in the collecting bottles for more than 2 hours then many of the eggs will deteriorate because of a lack of oxygen. You should place collecting devices over mature colonies just prior to the predicted spawning time and you need to monitor them at least once every hour. They should be removed each night and replaced the next night until spawning occurs. You should avoid leaving nets on colonies overnight as they are likely to be damaged and may cause considerable stress to the coral colonies if they are left underwater for more than a few hours. It is worth noting that *in situ* collection can only be done for hermaphroditic species (Box 5.1) as collectors are not designed to capture sperm from male gonochoric colonies.

Hermaphroditic corals can 'self-fertilise', however fertilisation rates are generally poor, therefore it is important to collect gametes from at least two colonies so that sperm from one colony can fertilise eggs from a different colony, this is known as cross-fertilisation. The size of your broodstock (i.e. the number of coral colonies that you collect gametes from) will affect the chances of successful larval culture. Relying on a small number of broodstock colonies is a risky strategy because some individuals may not spawn synchronously or may be incompatible with the rest of the broodstock (e.g. there may be 'cryptic' species that have a similar morphology to your target species but are not reproductively compatible). We recommend that gametes are collected from a minimum of 6 colonies (this also applies to method 3 below) for each species to ensure that sufficient numbers of gametes are obtained and to increase the amount of cross fertilisation. It is common for coral populations to 'stagger' spawning over several nights so having several colonies will improve the chances that at least two colonies spawn at the same time. We advise that you mix gametes from three or more colonies in order to improve fertilisation success and increase the amount of genetic diversity of the offspring.

An example of an in situ *gamete collection device attached to a spawning* Acropora *colony in Akajima, Japan.*
Above left: *Traps should be attached to a mature colony prior to spawning.*
Above right: *When the coral spawns, some of the gamete bundles are trapped in the jar.*
Left: *After spawning has finished, trapped gamete bundles can be collected (K. Vicentuan).*

3. Collecting from coral colonies *ex situ*

Corals can be removed from the reef by detaching the colony at the base using a 2 kg hammer and cold chisel. For *Acropora* corals removal should be relatively easy because the colony base is narrow, whereas massive or encrusting colonies may be more difficult to remove if they are firmly attached to the reef. For such colonies however it is usually possible to find healthy colonies with eroded bases that can be removed easily with little damage to the colony. For large colonies (i.e. >50 cm diameter) or encrusting colonies it may be better to remove a colony fragment. Colonies only a few centimetres in diameter may be sexually mature (see Box 5.1), however for larval rearing work it is beneficial to use larger colonies where possible. This will increase the number of gametes available for rearing. Ideally each colony or fragment should be 20 to 50 cm in diameter, although for some species only smaller colonies may be available.

If colonies are to be brought to a land-based hatchery for *ex situ* spawning, you need to transport them carefully in covered tubs or buckets filled with seawater. Large plastic coolers (>50 litres volume) with lids are ideal for transporting corals as they help to prevent water temperatures increasing during transportation. One cautionary note when using containers such as coolers with opaque lids is that closing the lid on a live coral simulates sunset. Consequently very mature corals transported in this way may experience enough darkness to spawn prematurely. This behaviour has been long observed and is a mechanism that can be used to bring forward spawning times by a few hours if desired. However, unless such an outcome is desired, it is best avoided by opening the lid every 15 minutes or so during transport to expose the colonies to daylight. If colonies are being transported by boat, travel times should be minimised, to not more than one hour when possible. If longer travel times are necessary, colonies should be given aeration using a portable battery operated aerator, or seawater in the coolers should be changed occasionally.

Prior to the night of spawning, you should maintain the colonies in the sea near the laboratory or in large fibre-glass aquarium tanks with sufficient water flow to exchange total tank volume several times each day. We recommend that tanks are large enough to provide 100 to 200 litres of water for each coral colony of approximately 30 to 50 cm diameter (i.e. 6 colonies can be kept in a 600 to 1200 litre tank with several daily water changes and aeration). In addition to adequate aeration and water exchange, species with dense branching or tabulate morphologies benefit from additional water circulation over the entire colony surface. This can be achieved by raising colonies above the tank bottom on plastic mesh trays or crates (to promote water circulation underneath the colony) and the addition of recirculation pumps to the tanks.

It is important to decide early on how you will mix gametes

from different colonies. There are basically three strategies that can be used. The simplest method is to keep all colonies from the same species together in a single tank during spawning so that gamete bundles from several colonies will be mixed at the water surface if and when colonies spawn together. Another method is to separate colonies into two containers. This is advisable when there are six or more donor colonies as it ensures against a single colony with non-viable eggs (e.g. an unhealthy colony or a cryptic species with incompatible eggs) compromising the entire culture. A final strategy is to isolate individual colonies in separate containers until they spawn and selectively cross fertilise various initial combinations and only later combine the most successful crosses for ongoing culture. This final method is more labour intensive than the first two methods, however it gives you much greater control and may be the best strategy when working with gonochoric species as it will allow you to determine the sex of individual colonies.

Provided colonies are in good health and oxygen levels in the water are adequate (6–8 mg/L – this can be checked with a handheld dissolved oxygen meter), water flow in the tanks should be turned off at around sunset each evening and colonies should be observed about every 30 minutes for signs of spawning. If you have placed colonies in small buckets or isolated tanks it is prudent to carry out partial water exchanges periodically to maintain water temperatures close to ambient. Many species spawn within three hours of sunset (Box 5.2), however the precise time of spawning will vary depending on location and species. Lights should be switched off in the hatchery and corals should be shielded from any artificial light during the monitoring period. For many hermaphroditic broadcasting species you will see egg bundles 'setting' in the polyp mouths 1 to 3 hours before spawning, for other species the presence of eggs at the surface or in the water column indicates spawning has occurred.

Once a colony begins to spawn, sperm, eggs or gamete bundles are released from polyp mouths. For hermaphroditic species, bundles are usually buoyant and will slowly rise to the water surface. You should allow all colonies to spawn completely before attempting to harvest the gametes. Spawning of colonies of the same species placed in separate tanks usually occurs within 15–30 minutes of one another. Once several colonies have completed spawning there should be a slick of buoyant gamete bundles containing eggs and sperm on the surface (for hermaphroditic species) and harvesting should begin. You can collect the bundles using a clean scoop (a plastic cup or bowl). It is important to collect as many bundles and as little water as possible in each scoop. This can be achieved by holding the lip of the scoop just under the water meniscus allowing gamete bundles to flow into the scoop. You then need to transfer the bundles to a clean container (usually plastic or polycarbonate) of known volume (e.g. 50 or 100 litres) about one third to half full with clean filtered sea water where fertilisation can be optimised (section 5.6).

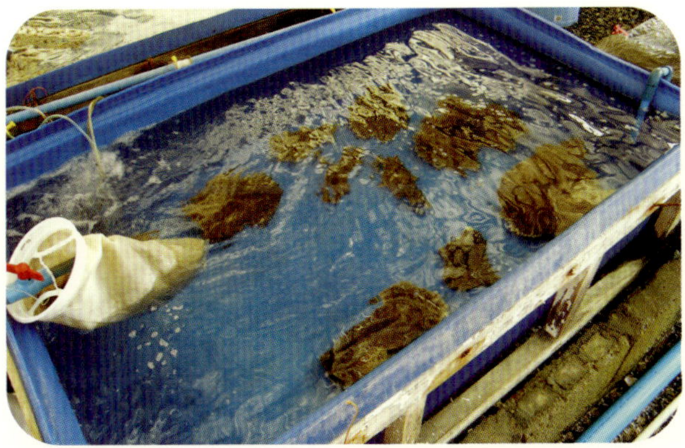

Coral colonies being maintained in a flow through land based hatchery tank (J. Guest).

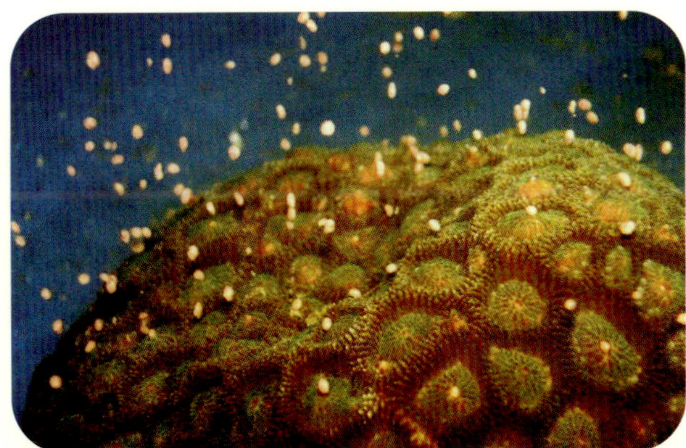

A faviid coral colony spawning in a tank in the Philippines (K. Vicentuan).

Eggs setting just prior to spawning in an Acropora colony (A. Heyward).

A colony of Acropora digitifera that has been isolated and has spawned in a plastic tub in Palau (J. Guest).

Left: A cup is used to collect the buoyant gamete bundles from just below the water meniscus so that bundles are concentrated (J. Guest).

Far left: Gametes being collected from the surface of a tank after several Acropora colonies have spawned together in Palau (J. Guest).

5.6 How to optimise fertilisation in broadcast spawning species

No matter what method is used to harvest gametes, the methods for fertilisation, maintaining embryos, and rearing the larvae until they are ready to settle are similar. Once you have collected all gamete bundles (methods 2 or 3 above) and transferred them to a separate container, the level in the container should be topped up to a known volume (100 litres makes subsequent calculations easier). Fertilisation will take place in this tank, therefore it is important that clean, filtered seawater is used (Box 5.3) and that the temperature is close to ambient seawater temperature. Temperatures higher than 4°C above normal ambient temperature markedly reduce fertilisation success in some *Acropora* species and we recommend that temperature does not deviate more than 1°C from ambient.

Gamete bundles will break apart and the water should become white and almost opaque as a result of the released sperm. It is generally accepted that sperm densities between 10^6 to 10^7 sperm per ml are most suitable. In practice sperm density seems to have little effect on fertilisation rates providing that gamete bundles from several colonies are mixed in a relatively small volume container (e.g. 50 to 100 litres). Egg density at this stage is also not critical. More than 1 million eggs can be fertilized in a single 100 L polycarbonate tank. If colonies have been isolated or divided into separate tanks then gametes from the separate batches should also be fertilised in separate fertilisation tanks.

As soon as the all of the gamete bundles have been transferred to the fertilisation tank(s) and the bundles have broken apart, you need to estimate the number of eggs you have. This is necessary to determine the volume of water required for stocking during the larval rearing phase (see below). Eggs are usually very buoyant so to estimate egg density, it is important to distribute the fertilising eggs in the container as evenly as possible by vigorously stirring using a large plastic spoon, paddle or plunger. One person should constantly stir while another takes several small samples from different parts of the tank using plastic screw top sampling tubes. You should take at least five 15-ml samples from a 100 litre tank to get an average density estimate and count the eggs immediately under a stereo microscope. Using the protocol in Example 5.1 (right), you can then estimate of the total number of eggs.

Significant numbers of eggs are fertilised within the first few minutes and fertilisation increases steadily for up to an hour after gametes are mixed together. Therefore the cross-fertilised gamete mixture should be left for between 15 to 30 minutes (not longer than one hour) with gentle stirring approximately every 5 minutes to prevent oxygen depletion. Once fertilisation has occurred it is important to 'clean' the eggs to remove excess sperm. Excessive numbers of sperm can lead to a problem known as polyspermy (where many sperm attempt to fertilise one egg). Furthermore, breakdown of excess sperm in the fertilisation or rearing tank will lead to a reduction in water quality and high mortality of embryos. Cleaning can be achieved by draining the fertilisation tank onto a plankton mesh sieve (100 µm) from the bottom via either a plug or a tube siphon and carefully re-filling with clean sea water several times. An alternative and convenient method if draining is not possible is to scoop eggs from the surface of the fertilisation tank and transfer them to another similar sized tank containing clean filtered sea water using the same method as used to scoop from the spawning tank (i.e. a clean cup is held just below the water meniscus to collect as many eggs and as little water as possible).

After the excess sperm have been removed, you need to gently transfer the eggs to a rearing tank at a suitable stocking density (see how to calculate in Example 5.2). After about 2 hours you should take a sample of the culture (approx. 50 ml) and check under a stereo-microscope for evidence of fertilisation (Figure 5.4). If sufficient numbers of fertilised embryos are present (>80% of eggs are dividing), the initial stages of larval rearing have been successful and you can proceed to rear the embryos through to larvae as described in section 5.8. It is important that there are not many unfertilized eggs in the culture as these will rapidly deteriorate and damage the water quality, killing other normal embryos. If fertilisation levels are low (<60%) it may be worth considering abandoning the culture. The fertilisation work should be completed within four hours of spawning.

Example 5.1 Estimating eggs numbers

Sub-sample (15 ml)	1	2	3	4	5	Average
Number of eggs per 15 ml	100	200	170	130	150	150 eggs per 15 ml
Number of eggs per 1 ml	6.7	13.3	11.3	8.7	10.0	10.0 eggs per 1 ml

Total number of eggs = 10.0 x 100,000[†] = 1 million
[†]for a 100 litre fertilisation tank

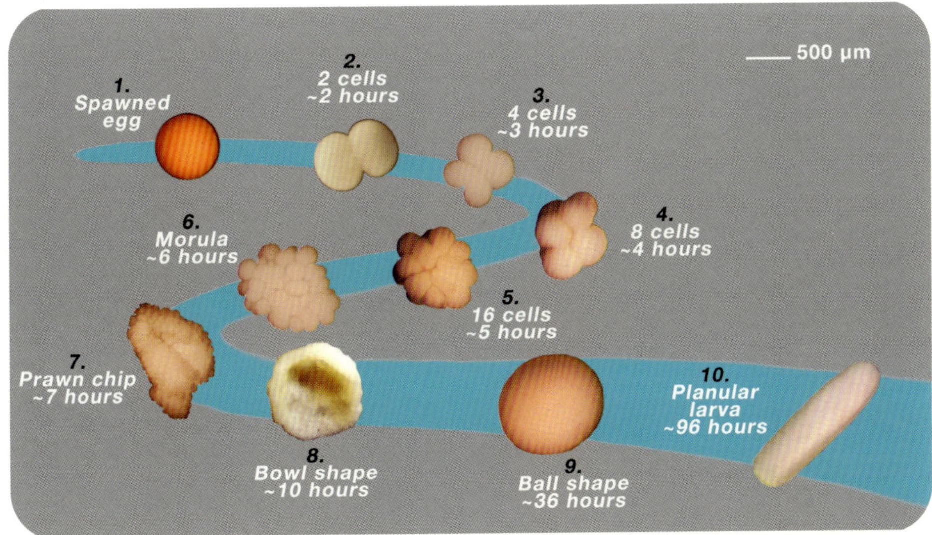

Figure 5.4. Embryo developmental stages for a typical *Acropora* coral (based on Okubo and Motokawa 2007[10]). 1) An unfertilised egg (or oocyte) with distinct spherical shape and opaque appearance; 2) about 2 hours after fertilisation occurs, embryos divide into two cells (known as blastomeres) by a process called cleavage; 3) at about 3 hours embryos divide again to become four cells; 4) cell division continues to produce 8 cells at about 4 hours; 5) 16 cells at about 5 hours and until the embryo consists of many cells 6) this stage is known as the morula and occurs after about 6 hours; 7) after about 7 hours the embryo is known as a 'prawn chip' because of its distinctive shape; 8) then after about 10 hours the embryo has a distinctive bowl shape; 9) and by 36 hours the embryo becomes completely 'ball shaped' and at this stage is considerably more robust and can withstand water changes (see section 5.8); 10) larvae gradually become more motile and elongated in shape until a 'cigar shaped' larva is formed at around 96 hours. (Photos 1–5, 7 and 9: C. Boch (*A. digitifera*); Photos 6 and 8: N. Okubo; Photo 10: M. Omori (*A. tenuis*).)

Example 5.2 Estimating suitable stocking density

Tank volume (litres)	100	500	1000	1500	2000
Min. tank surface area (m²)	0.08	0.38	0.75	1.13	1.50
Min. tank diameter (cm)[†]	31	69	98	120	138
Total number of propagules (eggs, embryos or larvae)[††]	30,000	150,000	300,000	450,000	600,000

[†]Minimum diameter is for a circular tank.
[††]Stocking density of 300 propagules per 1 litre volume or 40 propagules per 1 cm² surface area of sea water.

A 100 litre polycarbonate fertilisation tank containing eggs, sperm and embryos (J. Guest).

Establishing a suitable stocking density

It is essential that you stock the embryos at the correct density. If kept at too high a density there will be high mortality. The optimal volume will depend on various factors such as water temperature, water quality and frequency of water changes. Tank surface area is also important in the early stages of larval rearing because buoyant eggs will tend to aggregate at the surface of the tank. High survival of larvae can be attained with relatively little effort when embryos are kept at a stocking density of 300 propagules per 1 litre volume and 40 propagules per 1 cm² surface area of seawater (covering about 10–20% of the water surface area). You can make rearing tanks of any suitable material, e.g. fibre-glass, polycarbonate or PVC. If the work is being done on a budget or in remote locations, any available uncontaminated water-tight container can be used, e.g. inflatable paddling pools which are available from department stores or toy shops.

A 100 litre polycarbonate fertilisation tank containing gamete bundles being filled to a known volume. Notice the cloudiness of the water which is from the high sperm density (J. Guest).

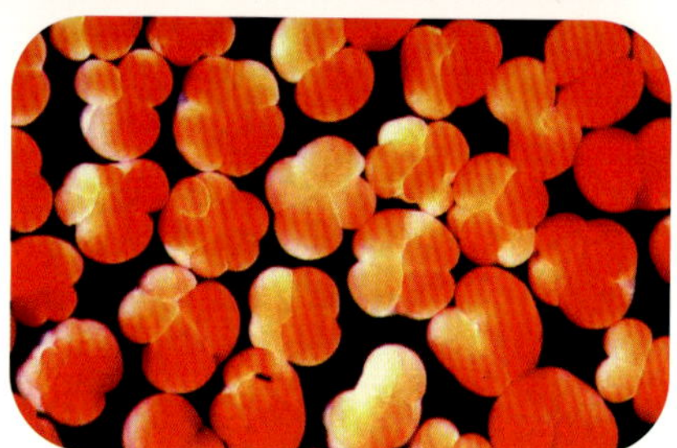

Above left: *Mixing gametes in the fertilisation tank so that eggs are evenly distributed to allow representative samples to be taken for counting (J. Guest).*

Above: *Pouring embryos into a large rearing tank about 1 hour after fertilisation (J. Guest).*

Left: *Early cleavage of eggs from an Acropora coral. This is approximately 2 hours after gametes have been mixed and many embryos are at two and four cell stages (A. Heyward).*

5.7 Collecting and rearing larvae from brooding corals

Rearing larvae from brooding corals is less complicated than rearing those from broadcasters because in most cases the planular larvae are already fully developed and ready to settle. You can harvest brooded larvae either by placing collecting devices over colonies *in situ* or by bringing colonies or colony fragments in to land-based aquarium tanks before the predicted time of planulation.

In situ *collection of planular larvae*

Collecting devices for *in situ* collection of planulae from brooding colonies consist of a plankton net with a 100–200 µm mesh size (plankton mesh can be bought from aquaculture and fishing supply stores) with an upturned plastic collection bottle attached at the mouth of the net. An upturned funnel can be inserted in the mouth of the collection bottle so that larvae can enter the bottle but not escape back down into the net. A plankton mesh covered window can also be cut in the top of the collection bottle to prevent water becoming stagnant in the bottle if left overnight. The net can be secured to colonies by means of a draw-string around the base of the net. Brooding corals tend to release planulae over an extended period of several hours, therefore you may need to leave nets attached to colonies overnight. For species that planulate during the night, you should deploy the larval collecting devices just before sunset and check them as early as possible the following morning. This method has been shown to be successful as a method for collecting larvae of *Stylophora pistillata* in Eilat[11]. However in some locations, *in situ* collection may not be feasible. After collection, larvae should be transported quickly to land and transferred to sea water tanks for settlement.

Ex situ *collection of planular larvae*

For *ex situ* collection, you should remove colonies from the reef and place in flow-through aquarium tanks (as described for broadcasting corals in section 5.5 method 3). There are several methods that can be used to collect released planular larvae. Colonies can be placed in a container inside the aquarium tank (e.g. a PVC cylinder or a plastic bucket) that has large plankton mesh windows inserted on the sides. You should use plankton mesh of a size that is small enough to trap larvae, but large enough not to become clogged by sediment and coral mucus (typically 100–300 μm mesh), and you should make the windows large enough to prevent rapid clogging. If the mesh becomes clogged then water will overflow into the main aquarium tank and larvae will be lost. The water level in the tank should always be below the top of the container allowing water to flow past the colony but trap any released planula larvae inside the container. Using this method it should be possible to maintain constant water flow in the tank, although it may be advisable to reduce water flow at night if that is when planula release peaks. Alternatively, corals can be held in tanks that have an overflow pipe to allow larvae to flow out of the holding tank and into a mesh-lined planula collector that is kept in a separate smaller tank. You can make the collector using two pieces of PVC pipe and a piece of plankton mesh. You glue the pipes together with the plankton mesh net placed in between the two pipes. The top of the collector should be above the top of the tank so that water flows out of the tank but larvae are trapped in the collector. We have used a PVC collector with a diameter of 20 cm, a height of 30 cm with 300 μm mesh plankton net and found this to be successful when collecting larvae from *Pocillopora damicornis*. Each morning and periodically through the day you should monitor containers for the presence of planular larvae and any found should be removed with a pipette, a siphon hose or other suitable collection device (e.g. plastic cup) and transferred either to clean holding tanks or to tanks containing substrates for settlement.

In situ *larval collection devices placed over colonies of Stylophora pistillata in Eilat (B. Linden/B. Rinkevich Lab.).*

Ex situ *larval collection using an overflow pipe and mesh collector (J. Guest).*

5.8 How to maintain embryos and larvae until they are ready to settle

Maintaining larvae from brooding species

For brooded larvae, settlement may begin soon after release, so potentially larvae can immediately be introduced to conditioned settlement substrata or areas of reef if direct enhancement is being carried out (see section 5.9). However for some Atlantic brooding species it has been shown that it is beneficial to maintain larvae in a separate clean tank for at least two days before attempting to settle them[12]. This is because, for certain species, at time of release there is a significant amount of both lipids and mucus associated with the larvae. It is necessary to wash away the mucus and lipids by doing regular water changes otherwise they will become a source of energy for unwanted bacteria in the settlement tanks. During this period, collected larvae should be maintained in clean filtered seawater at ambient temperature at densities of not more than 300 larvae per litre of seawater, with water changes done at least once each day (see below for water change techniques). Conversely, for larvae of some common Indo-Pacific species (e.g. *Pocillopora damicornis*) settlement can occur very soon after release and larvae will settle readily on almost any surface. For this reason, when working with *P. damicornis* it is advisable to make sure that your holding tanks and settlement tanks are thoroughly cleaned to remove biofilms that will encourage unwanted larval settlement and to have your conditioned settlement substrates ready as soon as larvae are released.

As brooded larvae contain zooxanthellae you should provide enough light for photosynthesis. Shaded sunlight should be adequate, however if larval cultures are maintained outdoors it may be difficult to control temperature in the culture tanks while still providing sufficient light. For this reason it is advisable for you to maintain your culture of brooded larvae in a temperature controlled room (with temperatures set close to that of ambient sea water) with an artificial light source. Ideally you should use a high intensity actinic lamp with strong emissions in the short wavelength region of the spectrum, peaking at 420 nm (e.g. Coral Sun® Actinic 420 T5-HO).

Maintaining embryos and larvae from broadcasters

For broadcasting species, you need to maintain developing

embryos in the rearing tanks until swimming larvae have developed. During this period, which is usually around 2–5 days, it is essential that you maintain a healthy environment for the larvae. You should keep rearing tanks shaded using a net that reduces 40–60% of direct sunlight and protect from rain showers as sudden reductions in salinity will increase larval mortality. Checking and maintaining water temperature within a normal range is critical. Ideally, you should not allow temperatures to rise above normal ambient levels (which will vary from one location to another) and this can be done by increasing shading above the tank and carrying out regular water changes as necessary.

During the larval rearing phase you should check embryos and larvae at least daily by examination under a stereo dissecting microscope to assess health and status (see Figure 5.4). An important consideration at this stage is handling of embryos. During the early stages, developing embryos must be treated extremely carefully so that dividing cells are not 'broken'. Rough handling of embryos during the cell division stage, i.e. from 1 hour post spawning until embryos are 'ball shaped' (see Figure 5.4), will result in many of the embryos not completing development or being smaller than normal. Once you have carefully transferred the embryos to the rearing tank you should leave them in static water without aeration until embryos have a rounded 'ball shape' (usually 24–36 hours after fertilisation). Gentle aeration can be introduced after this and should be gradually increased each day as larvae become more robust.

You should check water quality daily and carry out water changes after 24 hours if water quality has deteriorated or temperature has risen in the tank. Signs that water quality has deteriorated include cloudiness and the appearance of white foam on the water surface. Foam and floating scum can be removed using polyethylene plastic wrap (e.g. Glad wrap™ or Saran wrap™) by placing sheets on the water surface and allowing any scum to stick to the wrap before removing. However, in most cases such treatments are unable to reduce mortality once water quality has deteriorated due to bacterial propagation.

During water changes, you should treat embryos and larvae as gently as possible. You can do water changes by siphoning water from the rearing tank onto a submerged 100 µm mesh sieve or net. Larvae that are trapped on the sieve or net should remain submerged as the tank is being emptied and should be carefully replaced by 'backwashing' them into the tank. You can also use a two-sided sieve (see right) to do water changes. With this method, the sieve is placed in the tank so that embryos or larvae remain in the tank during the water exchange. Once the rearing tank is half empty it should be topped up with clean filtered sea water. Using these methods it is possible to do partial or full water changes.

Rearing tanks for embryos and larvae are typically kept on land; however it is possible to rear larvae in floating ponds

at protected sites close to the shore. This method has the advantage that temperature can be buffered by the surrounding sea water, furthermore water exchange can occur if mesh windows are built in to a floating tank. However, floating ponds are subject to inclement weather. If mesh screens are not incorporated for water exchange then it is necessary to cover the ponds to keep off rain water that will change the surface salinity. You will also need to clean mesh screens using scuba to prevent clogging by fouling organisms and sediment. In Okinawa, Japan, floating ponds have been used to rear coral larvae successfully[1, 13-14]. The ponds are constructed of vinyl sheet and are connected by floating rafts. Water is sprayed against the walls of the pond by means of a hose attached to the upper part of the pond, with holes made at intervals, supplied by a submerged pump. This prevents larvae sticking to the walls of the pond during rearing and promotes water exchange.

During the rearing phase it is necessary to assess larval health and readiness of the larvae to settle (see Box 5.4).

Large inflatable pools (4000 litres) being used as rearing tanks in Palau. Note that rearing pools are covered by a roof and shade netting to protect embryos from excessive sunlight and rain (J. Guest).

A floating rearing pond containing fertilised embryos in Akajima, Japan (M. Omori).

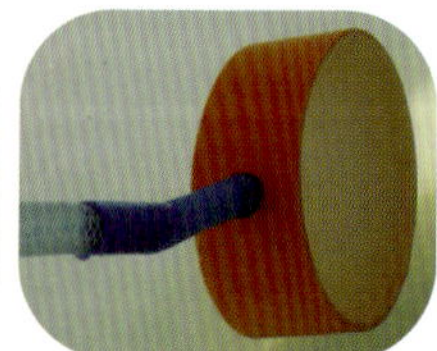

An example of a two sided sieve that can be used for convenient water changes (J. Guest).

Floating ponds used for larval rearing in Akajima, Japan (M.Omori).

Box 5.4 **Assessing settlement competency of coral larvae from broadcast spawners**

The time needed for the larvae to develop to a stage when they are ready to settle varies considerably among species and locations and is dependent on environmental factors such as temperature; therefore it is important that you visually assess levels of settlement competency during the larval development period. An effective method is to track a sub-sample of developing larvae held in smaller volume containers. You should remove approximately 4 litres of seawater containing larvae from the rearing tanks immediately after stocking. This sub-sample can be kept in clean plastic bottles (e.g. four 1 litre bottles) which can be left floating in the rearing tank (to buffer temperature). You should take samples (~400 ml) of larvae from one of these bottles at 12 hours post fertilisation by pouring into a clean cup; repeat this at 24 hours and subsequently every 24 hours until high levels (>80%) of larvae are ready to settle.

From the cup containing the sample of larvae, place approximately 20 larvae into each of 6 sterile replicate plastic wells or medicine cups containing about 10–20 ml of UV-treated, 0.2 μm filtered seawater. Plastic laboratory 6-well culture plates are ideal, however disposable plastic cups can also be used. You should flush the larvae well with clean seawater (ideally 0.2 μm filtered) before adding them to the culture wells. In seawater alone, settlement rates of larvae will be very low, therefore it is necessary to add an inducer for settlement and metamorphosis. The presence of certain species of crustose coralline algae (CCA), particularly *Hydrolithon* spp. and *Peyssonnelia* spp., has been shown to induce metamorphosis in a number of coral species[15-17].

A chip of CCA approximately 5 mm x 5 mm in size should be scraped from the surface of a larger piece of CCA that you have collected from the reef. You should thoroughly clean the chips using a soft brush while flushing with filtered seawater. Place one CCA chip in each well or cup. Alternatively, you can use a small piece of coral rock or dead coral that has been immersed in seawater for more than 2 months (so that crustose coralline algae is attached). You should keep the wells or cups indoors on a stable surface free from vibrations where temperature throughout the day remains within the normal range of ambient seawater temperature for the locality (i.e. not in an air conditioned room or close to working machinery). You should check each well or cup under the dissecting microscope after 24 hours and count the number of larvae in one of the following four conditions: 1) attached, 2) metamorphosed, 3) alive but not attached, 4) dead. Larvae are recorded as settled when they are either attached to the substrate or have metamorphosed into a polyp. The average percentage settlement can then be calculated and plotted on a graph (Figure 5.5). When the average settlement rate reaches at least 80% then you should introduce larvae from the main holding facility to settlement substrata for subsequent rearing or to areas of degraded reef. If only low levels of attachment and metamorphosis (e.g. <50%) are ever achieved, this may indicate that larvae are not healthy due to poor water quality in the rearing tanks.

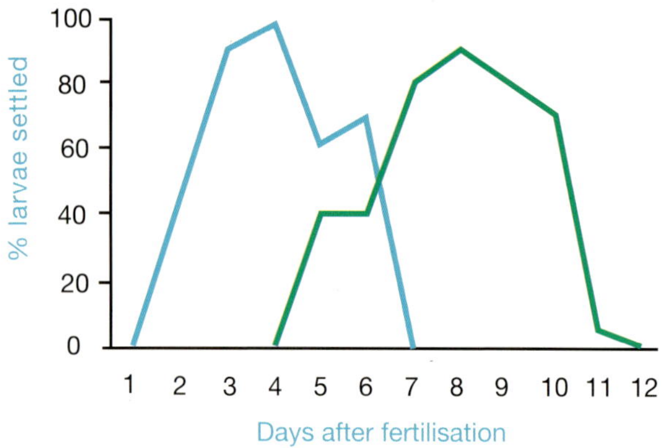

Figure 5.5. Typical competency curves for broadcasting species. The green line is for an *Acropora* (*A. millepora*) from Australia and the light blue for a faviid (*Favites halicora*) coral from the Philippines showing the average proportion (%) of larvae that settle (attach and metamorphose) in the days following fertilisation.

Assessing larval competency to settle. Small chips of crustose coralline algae (CCA) are placed in 12 ml culture wells and about 20 larvae are placed into each well with a disposable pipette (J. Guest).

Settled and metamorphosed larvae on a chip of crustose coralline algae (A. Morse).

5.9 How to settle coral larvae for reef restoration

Ways of establishing settled larvae in the field are still at the experimental stage, however there are essentially two methods which you can use to introduce larvae to an area of degraded reef: 1) you can settle larvae on to purpose-made substrates for nursery rearing prior to the subsequent transplantation of the substrate and surviving juvenile corals to the field, or 2) you can introduce competent larvae *directly* to areas of reef at high densities. Because of the very high rates of coral post-settlement mortality in nature, the second approach is less likely to assist your rehabilitation efforts. However, if the mortality of the newly settled larvae can be reduced sufficiently by the protection afforded by nursery or hatchery rearing, the first approach may offer considerable potential for coral reef restoration. This is a key area of research at present.

Settlement substrates for coral larvae

Techniques for mass producing larvae are reasonably well established, however the techniques for rearing these larvae until they can be transplanted to the reef in sufficiently large numbers are still experimental. Until recently, ceramic or terracotta tiles had been used for most experiments involving coral larval rearing and transplantation to reefs. However, while tiles are useful for experiments they are not the most suitable substrates to use for restoration efforts, because a) it is not easy to control where larvae settle or to separate settlers, b) tiles do not have a specific mechanism for attachment to the reef, and c) tiles cannot easily be handled without damaging settled corals.

Alternative substrates for larval settlement that you can use for restoration and that are currently being trialled typically have two parts: i) an area where coral larvae will preferentially settle and ii) a device that will allow both handling without causing damage to the corals and easy attachment to the reef. Various considerations should be taken into account when designing such substrates including, the size of each substrate, the cost and durability of the materials and use of a surface texture that will enhance settlement and survival of corals. Natural post settlement mortality of corals is likely to be high even in a nursery or hatchery. Ideally, one coral should survive on each substrate to a size where it can be transplanted. Currently however, we do not know how many corals should be settled initially on each substrate to maximise the chance of having one surviving transplantable individual per substrate. It may be beneficial if settlement substrates are not totally smooth, as grooves or crevices potentially provide refugia for small corals that may prevent them being removed by grazers or predators.

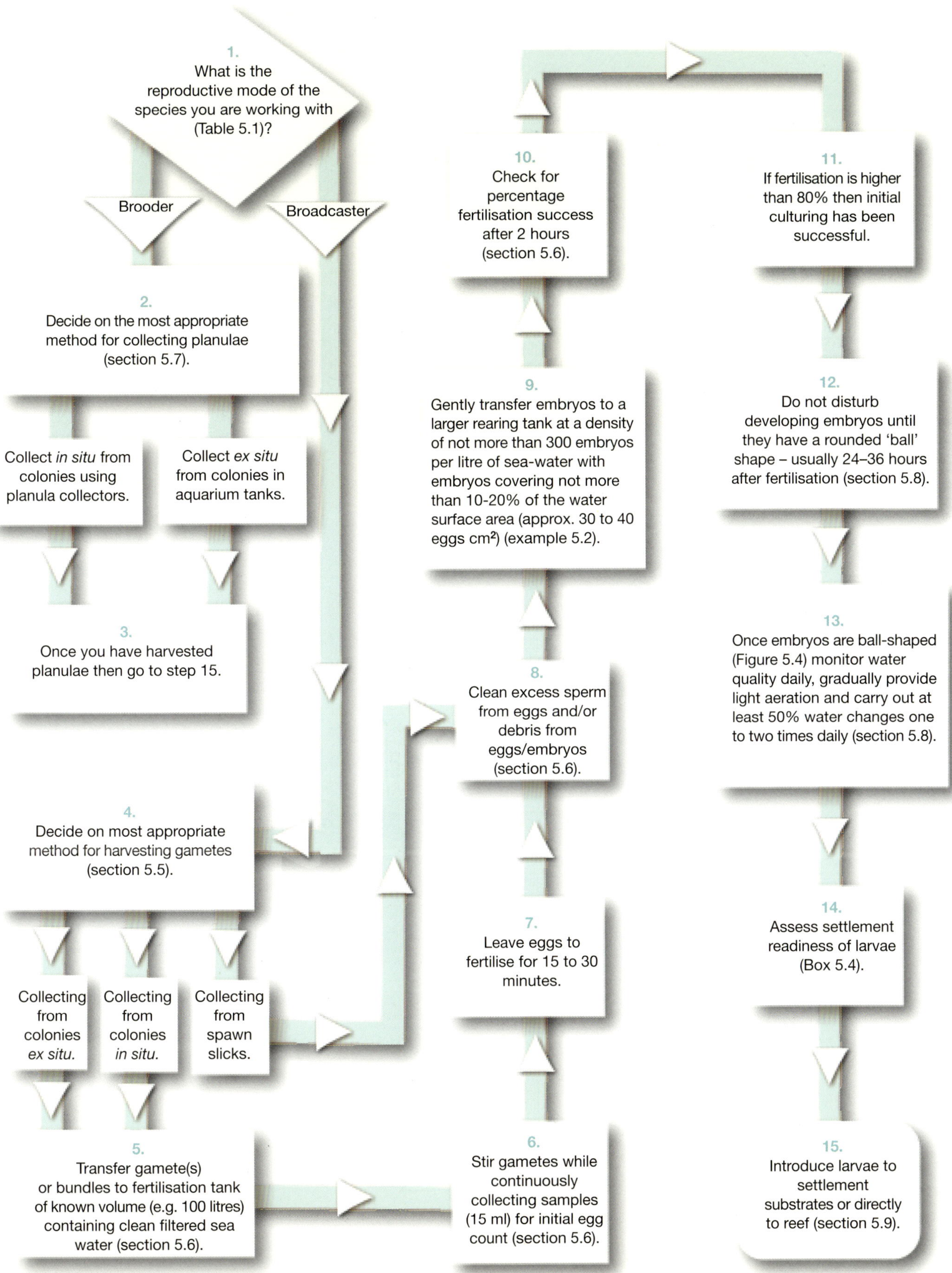

Figure 5.6. Steps in rearing larvae from brooding and broadcast spawning corals.

Below, we describe three examples of coral settlement substrates that are currently being trialled.

1) The Coral Settlement Device (CSD)[9] is being used in Sekisei Lagoon (Okinawa, Japan). The CSD is made of ceramic and has a diameter of 44 mm a height of 34 mm. It is composed of a disc (upper diameter 44 mm, underside diameter 41 mm, height 11 mm), a spacer (upper diameter 24 mm, underside diameter 22 mm, height 10 mm) and a leg (upper diameter 10 mm, height 13 mm). There are eight radial grooves on the underside of the disc (4 mm in width and depth). There is a hollow on the upper side of each disc so the leg of a CSD can be inserted to form a 10-mm-high vertical space between discs and allow CSDs to be stacked one on top of another (see Ch. 8: Case study 7).

2) The Coral Peg[18] is being trialled in Akajima (Okinawa, Japan). The 'head' or settlement area is made of cement mixed with quartz sand and has a diameter of 1.8 cm and height of 1.0 cm. The shaft is plastic and is 1 cm in diameter and 5 cm in height.

3) The Coral Plug-in is currently being trialled in Bolinao (northwest Philippines). The settlement area is cylindrical and made of cement (ratio of 1 part river sand to 1 part Portland cement) with a diameter of 2 cm and a height of 1.5 cm. The attachment part is a standard plastic wall plug (available from most hardware suppliers) and has a diameter of 1 cm and a length of 5 cm.

4) Commercially available masonry push mounts made of weather resistant nylon are being trialled as a substrate for settlement in Palau. Push mounts come in various sizes with the current ones being tested having a length of 3.8 cm and width of 1.4 cm (for the settlement part) and a diameter of 1.2 cm for the attachment part.

Whatever substrate is used it is essential that you 'condition' it for a period of time in seawater before attempting larval settlement (see Box 5.5). To settle reared larvae onto the substrates, you should place them in the rearing tank as soon as larvae have reached peak settlement competency (see Box 5.4). Settlement substrates can be suspended in the settlement tanks by strings (for narrow deep tanks) or can be arranged on the bottom of the tank (for wide shallow tanks). During the settlement period, moderate aeration should be maintained and you should carry out daily water changes. You should check each day for the presence of settled larvae and we recommend you do counts of a few randomly selected substrates every 24 hours to estimate settlement success. Depending on the type of substrate you are using, it might be advisable to introduce different settlement substrates (e.g. ceramic or terracotta tiles) that are easier to use for specifically estimating settlement success.

A coral rearing substrate in a settlement tank with newly settled coral spat (the red spots). This substrate has been conditioned for over 1 year and is encrusted with CCA – the larger purple patches (J. Guest).

The 'Coral Peg' used in Akajima, Japan with left: newly settled Acropora coral spat and right: a 14 month old juvenile coral (M. Omori).

The 'coral plug-in' used in Bolinao, Philippines consisting of a 10 mm plastic wall plug and concrete head. Arrows show newly settled Acropora spat.

Coral 'plug-in' at in situ nursery with a six-month old Acropora colony.

Transplanting a coral plug-in with a 14 month old Acropora colony from an in situ nursey onto the reef (J. Guest).

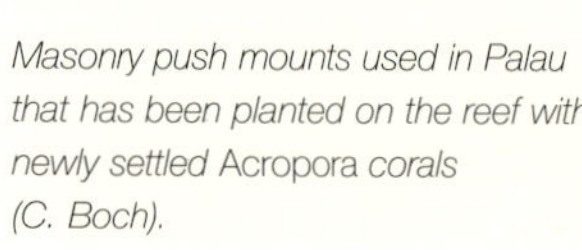

Masonry push mounts used in Palau that has been planted on the reef with newly settled Acropora corals (C. Boch).

Box 5.5 Conditioning substrates for coral settlement

Many marine organisms prefer to settle on substrates that have been 'conditioned' in sea water for a period of time. Conditioning broadly refers to the biological succession that occurs on any substrate that is submerged in seawater. Typically, this begins with micro-organisms and is followed by settlement of various types of algae and invertebrates. Corals settle preferentially on substrates that have had time to develop a biofilm and settlement is considerably enhanced when substrate surfaces have some crustose coralline algae (CCA). Conditioning of settlement substrates can be done in the sea or in flow-through aquarium tanks. Pieces of CCA harvested from the reef can be included in conditioning tanks to help speed up the conditioning process. The optimal length of time for conditioning is still unknown, however eight weeks of conditioning has been shown to significantly enhance settlement compared to substrates conditioned for two weeks (Figure 5.7). Little is known of the effects, if any, of conditioning on post-settlement survival. If other sessile organisms have grown on the substrates during conditioning, for example turf or macro-algae, zoanthids, sponges, bryozoans and ascidians, they should be removed by brushing before coral larvae are allowed to settle.

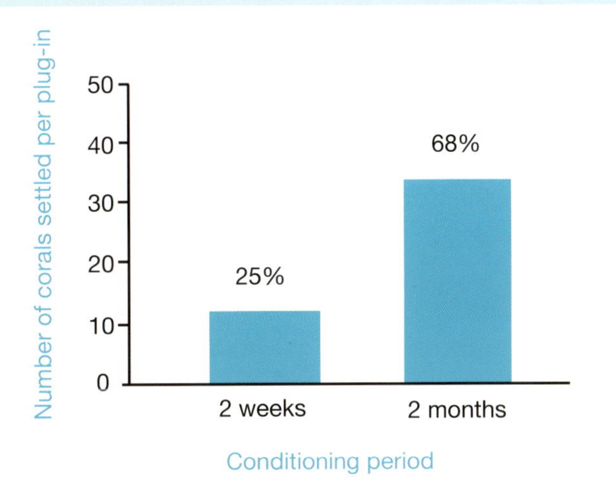

Figure 5.7. The effect of different conditioning periods on settlement success of *Favites halicora* coral larvae in the Philippines. The graph shows the difference in average number of corals settled on individual coral plug-ins and average percentage settlement. Each coral plug-in has an area of 14.9 cm^2. Plug-ins were conditioned for c. 2 week and c. 2 months and approx. 50 larvae were introduced to each plug-in.

How to rear corals from settled spat to juveniles

After sufficient numbers of corals have settled (usually this happens between 1 and 5 days after larvae become competent), you need to either transfer substrates to a) areas of degraded reef, b) flow-through aquarium tanks for a period of *ex situ* rearing, or c) a protected field nursery for a period of *in situ* nursery rearing. If corals are transplanted directly to the reef immediately after settlement, it is likely that survival levels will be very low. Maintaining the corals in *ex situ* tanks for a period of time after settlement can significantly increase survivorship because water quality can be controlled and grazers and predators are not present. However, keeping juvenile corals in aquarium tanks significantly increases labour and costs. As an alternative to land-based tank rearing, corals may be transferred to *in situ* nurseries. Typically, these are located in shallow areas with good water quality and protection from storms (e.g. in lagoons). In recent years there have been considerable advances in methods for rearing of asexual coral fragments at *in situ* nurseries (see Chapter 4), and these practices can also be adapted for rearing sexually produced corals.

There are several considerations to be made when setting up nurseries in the field. These include the level of maintenance likely to be needed, how to deal with biological fouling (e.g. from algae, sponges, sea squirts, molluscs, etc.), water quality, sedimentation, diseases, predation, etc. When corals are very small they may be removed by grazing fish and this can be a significant source of mortality. However, grazers have an important role in removing algae that may potentially smother and kill young corals. In Okinawa and Palau, juvenile corals cultured from eggs have been successfully maintained in mid-water cages in a co-culture with the grazing snail *Trochus niloticus* [4,19]. The caging prevents removal of corals by grazing fish, while the snails graze the algae without causing damage to the corals.

It is important to decide the optimal length of time to rear spat *ex situ* in tanks or at an *in situ* nursery site before corals are transplanted directly to a degraded reef. Very little information exists on the optimal size or the extent of survival that can be achieved by varying the rearing times *ex situ* and *in situ*. In the Philippines, juvenile colonies of *Pocillopora damicornis* that were reared *ex situ* to at least 10 mm in diameter had approximately 50% survival 12 months after transplantation[20]. Whereas in Okinawa, after a series of trials, it was found that transplantation of *Acropora* colonies of about 6 cm in diameter after 1.5 years of *in situ* cage culture was the most successful strategy. 2000 corals at this size class were transplanted to the reef and after six months almost 90% were still alive[21]. Unfortunately a major typhoon destroyed many of the transplanted corals leaving only ~160 (8%) alive 2.5 years after transplantation. Some of these colonies spawned naturally for the first time when they were four years old.

How to 'introduce' coral larvae directly to the reef

Although trials indicate that direct enhancement with *Acropora* larvae has no significant effect on recruit density within a year, the techniques are described in case others wish to test the method at specific locations or with other genera. Direct enhancement should be done when larvae are at the peak of competency and this time will vary depending on species and location (see Box 5.4). Various techniques can be used for direct seeding at various scales and need to be adapted to local circumstances. A technique that combined floating larval rearing ponds with direct enhancement has been used in Western Australia[2]. This method involved rearing embryos collected from natural spawn slicks in an anchored floating pond, then pumping competent larvae directly onto an area of natural reef covered by mesh enclosures via hoses connected to the bottom of the pond (Figure 5.8). Another technique that has been trialled in Palau (Micronesia) involved introducing competent larvae into inner-mesh camping tent screens that had been placed over experimental artificial reefs. The tent base was reinforced and weighted down with metal 're-bar' and lengths of rubber hose pipe were placed over the tent frame to make it flexible but durable. Competent larvae of *Acropora digitifera* (total of approx. 1 million larvae) that had been reared in tanks were transported in plastic coolers (100 litres) and poured directly in to each of seven tents from the deck of a boat by connecting a length of flexible plastic hose to a valve on the top of each tent. Tents were left over each artificial reef for 24 hours before being carefully removed. In Okinawa a similar technique was tested[13] where 1.6 million planula larvae reared from slicks were cultured in floating ponds and released over concrete blocks surrounded by a nylon net enclosure. A soft polyethylene container was used for transportation of the larvae. These were then released into the enclosures by a scuba diver inverting the container underwater.

Trials using these methods have shown that early recruitment can be significantly enhanced but have failed to show any long term effects on the numbers of surviving corals on enhanced substrates after 12 months. It seems that even if early recruitment is significantly increased, the majority of these settled corals die within a few months due to natural processes. We recommend that this method is not adopted as a reef restoration technique unless positive evidence of a long term effect on coral reef recovery is forthcoming.

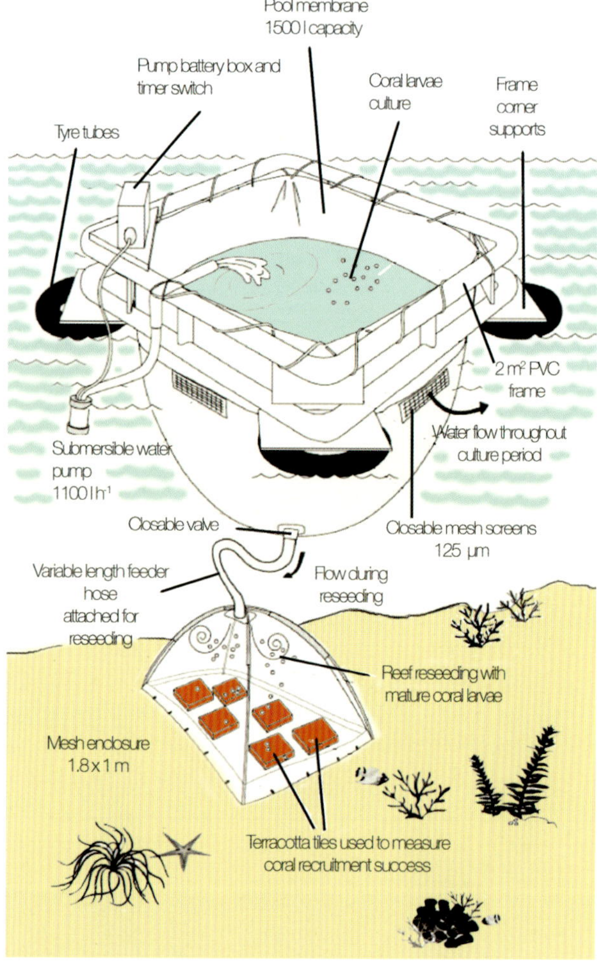

Figure 5.8. Schematic of floating larval culture pond and reseeding system from Heyward et al. (2002).[2]

Using a mesh tent to directly seed larvae onto an artificial reef structure in Palau (J. Guest).

Table 5.3 Where to obtain equipment needed for rearing coral larvae.

Use	Item	Where to obtain
Assessing reproductive status	Chisel Hammer Screwdriver	Hardware store
Transporting colonies	Plastic cooler boxes with lids (40–100 litres)	Hardware store
Collecting spawn *in situ*	Plankton nets (100–300 μm mesh) and bottles (1 litre)	Aquaculture / aquarium suppliers
Harvesting, fertilisation and rearing	Plastic scoops (assorted sizes) Plastic hose-pipe/tubing (various diameters) Assorted plastic buckets (15–20 litres)	Hardware store
	Graduated jugs and beakers (various sizes)	Scientific equipment supplier
	Filter bags (1–10 μm) and filter cartridges (0.2–100 μm) Plankton net (100–300 μm mesh) Aerators and air stones (assorted sizes) Fertilisation tanks (50–100 litre) Rearing tanks (size dependent on scale of rearing effort)	Aquaculture / aquarium suppliers
	Inflatable paddling pools (for large scale low-cost rearing; size dependent on scale of effort)	Toy / Department store
Assessment of settlement readiness and larval health	6-well plastic culture plates (well volume ~12–17 ml) or medicine cups Disposable pipettes Dissecting microscope (magnification up to at least 32 x, ideally 64 x) Hand-held tally counter	Scientific equipment supplier
Rearing of settled coral spat	Materials needed for nursery construction Floating larval rearing ponds Materials for coral settlement substrates (e.g. wall plugs, cement, push mounts, plastic pins, ceramic substrates, etc.) Plastic cages for rearing juvenile corals with *Trochus* Underwater compressed-air drill	Hardware store
All field work	Boat (make sure that size, engine, deck space available are adequate to needs)	Rent or buy (whichever more cost-effective
	Scuba diving equipment (as needed)	Scuba diving retailer

5

References

1. Hatta, M., Iwao, K., Taniguchi, H. and Omori, M. (2004). Restoration technology using sexual reproduction. Pp.14–33 in Omori, M. and Fujiwara, S. (eds.) *Manual for restoration and remediation of coral reefs*. Nature Conservation Bureau, Ministry of the Environment, Japan. [Download at: www.coremoc.go.jp/report/RSTR/RSTR 2004a.pdf]

2. Heyward, A.J., Smith, L.D., Rees, M. and Field, S.N. (2002) Enhancement of coral recruitment by *in situ* mass culture of coral larvae. *Marine Ecology Progress Series*, 230, 113-118. [Download at: www.int-res.com/articles/meps2002/230/ m230p113.pdf]

3. Petersen, D. and Tollrian, R. (2001) Methods to enhance sexual recruitment for restoration of damaged reefs. *Bulletin of Marine Science*, 69, 989-1000.

4. Omori, M. (2008) Coral reefs at risk: the role of Japanese science and technology for restoration. Pp. 401–406 in Leewis R.J. and Janse, M. (eds) *Advances in Coral Husbandry in Public Aquariums*. Public Aquarium Husbandry Series, vol. 2. Burgers' Zoo, Arnhem, The Netherlands.

5. Baird, A.H., Guest, J.R. and Willis, B.L. (2009) Systematic and biogeographical patterns in the reproductive biology of scleractinian corals. *Annual Review of Ecology, Evolution and Systematics*, 40, 551– 571.

6. Hall, V.R. and Hughes, T.P. (1996) Reproductive strategies of modular organisms: comparative studies of reef-building corals. *Ecology*, 77, 950-963

7. Veron, J.E.N. (2000) *Corals of the World*. Australian Institute of Marine Science, Townsville, Australia.

8. Fukami, H., Chen, C.A., Budd, A.F., Collins, A., Wallace, C., Chuang, Y.Y., Chen, C., Dai, C.F., Iwao, K., Sheppard, C. and Knowlton, N. (2008) Mitochondrial and nuclear genes suggest that stony corals are monophyletic but most families of stony corals are not (Order Scleractinia, Class Anthozoa, Phylum Cnidaria). *PLoS ONE*, 3, e3222, doi:10.1371/journal.pone.0003222 [Download at: www.plosone.org/article/info%3Adoi%2F10. 1371%2Fjournal.pone.0003222]

9. Okamoto, M., Nojima, S., Fujiwara, S. and Furushima, Y. (2008) Development of ceramic settlement devices for coral reef restoration using *in situ* sexual reproduction of corals. *Fisheries Science*, 74, 1245-1253.

10. Okubo, N. and Motokawa, T. (2007) Embryogenesis in the reef-building coral *Acropora* spp. *Zoological Science* 24: 1169-1177.

11. Amar, K.O., Chadwick, N.E. and Rinkevich, B. (2007) Coral planulae as dispersion vehicles: biological properties of larvae released early and late in the season. *Marine Ecology Progress Series*, 350, 71-78.

12. Raimondi, P. T. and Morse, A.N.C. (2000) The consequences of complex larval behavior in a coral. *Ecology*, 81, 3193-3211.

13. Omori, M., Aota, T., Watanuki, A. and Taniguchi, H. (2004) Development of coral reef restoration method by mass culture, transportation and settlement of coral larvae. Pp. 31–38 in Yukihira, H. (ed.) *Proceedings of Palau Coral Reef Conference*. PICRC Publication 04-001.

14. Omori, M., Shibata, S., Yokokawa, M., Aota, T., Watanuki, A. and Iwao, K. (2007) Survivorship and vertical distribution of coral embryos and planula larvae in floating rearing ponds. *Galaxea*, 8, 77-81.

15. Morse, D.E., Hooker, N., Morse, A.N.C. and Jensen R.A. (1988) Control of larval metamorphosis and recruitment in sympatric agariciid corals. *Journal of Experimental Marine Biology and Ecology*, 116, 193-217.

16. Morse, A.N.C., Iwao, K., Baba, M., Shimoike, K., Hayashibara, T. and Omori, M. (1996) An ancient chemosensory mechanism brings new life to coral reefs. *Biological Bulletin*, 191, 149-154.

17. Heyward, A.J. and Negri, A.P. (1999) Natural inducers for coral larval metamorphosis. *Coral Reefs* 18, 273-279.

18. Omori, M. and Iwao, K. (2009) A novel sustrate (The "Coral Peg") for deploying sexually propagated corals for reef restoration. *Galaxea*, 11, 39.

19. Omori, M. (2005) Success of mass culture of *Acropora* corals from egg to colony in open water. *Coral Reefs*, 24, 563.

20. Raymundo, L.J. and Maypa, A.P. (2004) Getting bigger faster: mediation of size-specific mortality via fusion in juvenile coral transplants. *Ecological Applications*, 14, 281-295.

21. Omori, M., Iwao, K. and Tamura, M. (2008) Growth of transplanted *Acropora tenuis* 2 years after egg culture. *Coral Reefs*, 27, 165.

Chapter 6.

Methods of coral transplantation

99

Edgardo Gomez, Rommi Dizon and Alasdair Edwards

6.1 Introduction

There is unfortunately a widespread perception that "reef restoration" is synonymous with coral transplantation and often also with the use of artificial reefs. This unbalanced focus on *active* restoration interventions rather than *passive* restoration via good management of reef resources is possibly a product of the relative newsworthiness of the two approaches and the shorter timescale over which visible results can be achieved. Hopefully, the preceding chapters have made it clear that active rehabilitation still remains a risky process which is unproven on any significant scale (e.g. tens of hectares)[1]. On the other hand, it is clear that huge areas of reef can recover naturally from devastating natural disturbances, such as the 1998 mass-mortality in the Indian and West Pacific Oceans, at locations where reefs are relatively unimpacted by humans and retain their resilience (e.g. Palau, Chagos, Maldives). Similarly, recovery of reefs from the 2004 tsunami appears to be progressing well in those areas of Thailand where reefs are under relatively little anthropogenic pressure. Thus, there is good evidence that passive restoration, in the right circumstances, can work at large scales. As such it is the option of first choice and coral transplantation still remains an approach of last resort[1-2]. Nevertheless in those instances where passive management measures have failed to achieve recovery, active restoration can play a crucial role in kick-starting recovery processes[3].

Transplantation can be a cost-effective option for small scale rehabilitation efforts that do not divert funding from other coastal management priorities. For example, transplantation of corals to patches of denuded reef close to diving resorts funded by paying guests, or repair of the reef at ship-grounding sites where there is funding available from damage compensation payments. Transplantation may also be necessary (again as a last resort) when decisions have been taken to go ahead with a development (port or other coastal construction, channel dredging, pipeline laying, etc.) that threatens reefs, such that corals will be killed unless moved to a safe location. Finally, in the face of global climate change there *may* be potential for using the methods of coral rearing outlined in Chapter 4 and 5 to propagate resistant strains which can be transplanted to selected priority sites.

Since coral transplantation was first proposed as a way to shorten the recovery time of denuded coral reefs in Hawaii[4], the majority of coral reef restoration studies have focused on methods of transplantation. Transplants used have ranged from whole coral colonies (usually in compensatory mitigation projects), fragments collected from donor colonies in the wild, unattached fragments rescued from the reef ("corals of opportunity"), fragments reared in nurseries until they have grown to become small colonies (Chapter 4), or small colonies reared from coral spat settled on substrates in hatcheries (Chapter 5). Coral transplantation remains the current method of choice in active reef restoration since it

results in an immediate increase in live coral cover and substrate complexity and thus attracts fish and invertebrates to the degraded area. Where significant numbers of herbivores are attracted then algal grazing increases which can benefit corals and other sessile invertebrates by creating space for larval settlement and preventing phase shifts to a system dominated by macroalgae. This chapter focuses on the methods used to transplant corals from either one reef area to another or from nurseries to degraded reef areas in need of rehabilitation.

6.2 Rationale for coral transplantation

Coral transplantation is just one option available to managers considering how to rehabilitate a reef. The decision as to whether or not you should attempt it should be dealt with at the project planning stage (Chapter 2) with due considerations of the risks involved (Chapter 3). If good cost estimates were available, you would ideally compare the costs and benefits of alternative management approaches (see Chapters 7 and 8) before proceeding, however, in reality there may be political, social, legislative or other non-ecological reasons which ultimately decide whether coral transplantation is attempted at a site. For example, in large areas that have suffered natural disasters such as coral bleaching, the only practical option may be to leave nature to take its course, but for a resort that has suffered from the same natural stress, the manager may decide to take immediate local action to accelerate the recovery of the damaged reef through coral transplantation. The key word is *accelerate*. Often the primary reason for coral transplantation appears to be human impatience with the speed of natural recovery processes. However, there may be cases where, without some active intervention, a reef that has suffered a disturbance will not recover and coral transplantation can kick-start recovery. The crucial prerequisite for coral transplantation is that any significant local anthropogenic impacts on the reef are under some form of effective management. Otherwise, there is a high risk that transplanted corals will not survive.

Threats to coral reefs and reasons for carrying out reef rehabilitation are discussed in our companion volume *Reef Restoration Concepts & Guidelines*[5]. Essentially, transplanting coral fragments by-passes the vulnerable larva-to-juvenile stage in the coral life-cycle (see Chapter 5) where maybe one in a hundred thousand to one in a million larvae survive to become juvenile corals a few centimetres across. Transplanting fragments results in an immediate increase in live coral cover. Corals provide most of the three-dimensional structure of the reef and thus increase the available niches for marine organisms, provide habitats and refugia for larvae and juveniles of various species, and create topographic complexity that alters the physico-chemical environment (e.g. nutrient cycle, light and water motion regimes) of an area – acting as "ecological engineers". Depending on the size of fragments transplanted, what might have taken 2–5 years by natural larval recruitment and colony growth can be *approximated* in one transplantation event. However, natural coral recruits have had to settle out of the plankton and then survive for several years and are thus likely to be well-adapted to a site, whereas transplants may not be, and indeed they may be subject to significant risk of mortality (Chapter 3) even in a well-planned project. Further, given the costs of *active* rehabilitation (Chapters 7 and 8), scarce funding may be more effectively spent on supporting local management of reefs by improving water quality and reducing overfishing with the aim of increasing their resilience and thus ability to support sustainable employment in coastal communities[1].

also be carried out for other reasons. Many nations with coral reefs rely heavily upon tourism for much of their economic input, and attractive healthy coral reefs draw more repeat visitors than degraded or destroyed ones. In some beach resorts, coral reefs are not very accessible, so coral communities are artificially assembled on either natural or artificial substrates to add aesthetic and tourism-related value to the underwater landscape (e.g. Ch. 8: Case study 3). However, high diver pressure can have negative effects on reefs. Therefore, transplants can also be used to create designated recreational areas or snorkel trails for tourists to frequent, often using artificial reefs and thereby theoretically reducing the impact of divers and snorkelers on natural

Porites cylindrica transplants attached to a degraded coral mound in a lagoon in Philippines with epoxy putty, 4 months (left) and 36 months (right) after transplantation (P. Cabaitan). For the purposes of this experimental transplantation only a single species was used.

A special case of transplantation is where it is being carried out to "rescue" corals that will otherwise be destroyed or severely stressed by a coastal development. In such cases a substantial number of coral colonies[6], which may be of large size[7], may need to be moved to a nearby area with hard substrate that is not necessarily a degraded reef. This can present a problem as natural areas of hard substrate that are relatively devoid of corals are probably so for some good reason (which may not be immediately apparent even to a trained biologist). If corals are not naturally present at a site, then conditions for coral survival are unlikely to be conducive at that site, thus the transplanted colonies may not survive well. Similarly, if the site already has a natural coral community, the additional colonies may be increasing the long-established natural density in a non-sustainable way. In a few cases, there may be suitable degraded reefs (i.e. ones where the environment is very similar to the reefs being lost) which would benefit from the transplants but this is likely to be fortuitous and cannot be planned. Thus attempts to move threatened coral communities have significant risks attached. A lack of suitable natural hard substrate may lead to the use of artificial structures (commonly made of concrete) as a substitute.

Thus for most rehabilitation projects, the problem is finding enough suitable transplants of coral species that will survive at the rehabilitation site, whereas for mitigation exercises, where corals are being relocated from an impacted site, the problem is finding a suitable site to which to move them.

Although coral transplantation is primarily concerned with damage mitigation and promotion of reef recovery, it may

reefs[8]. This approach has been tried in the Red Sea, Maldives and Thailand among other places. It is controversial as it will only work if the artificial structures do not attract more divers to an area, but just divert some of the existing diver pressure (particularly that of novice divers with poor buoyancy control) to the artificial reefs.

As emphasised in Chapter 1, transplantation should generally only be attempted at sites where local anthropogenic impacts are under some form of management control. Otherwise, your transplants are likely to be at significant risk. For example, one net fishing expedition at a transplant site can easily damage or detach hundreds of coral colonies. The level of protection can range from strong formal protection such as in a marine protected area (MPA), marine reserve or marine park to informal community oversight where there is good stakeholder involvement or tourist resorts where the local reef is under the resort operator's control.

A team of free divers participating in a community reef restoration project on a shallow, sheltered degraded reef in Santiago Island lagoon, north-west Luzon, Philippines (D. dela Cruz).

6.3 Sources of transplants

Sourcing corals for rearing in nurseries for reef rehabilitation is discussed in detail in section 4.4 and the advice given there applies equally to the sourcing of transplant material for direct transplantation. In most instances you should be seeking to minimize any collateral damage to the natural reef, in which case there are two main sources of transplants. Firstly, there are "corals of opportunity", which are fragments broken off from coral colonies by natural processes (e.g. storms and certain fishes) and human activity that can be found lying on the seabed where they are in danger of being moved around by waves, further broken, abraded or smothered by sediment. Such corals generally have a low chance of survival, except in those species (e.g. several species of *Acropora*) that naturally reproduce by fragmentation[9], and their probability of surviving can be significantly improved by attachment to the reef. There may also be individual coral colonies that are at risk – badly bio-eroded at the base, already detached, being overgrown (e.g. by algae or sponges), etc. These can be rescued, have dead or diseased parts removed, and either be reared in nurseries prior to transplantation, or directly transplanted to the rehabilitation site. Secondly, there are donor colonies from which fragments should always be excised with care, ensuring that the colony is not dislodged or left with large wounds that become colonized by algae or boring sponges. The impacts on reproduction and survival of removing parts of a coral colony have been studied in very few species. Research suggests that as a precautionary measure, you should only remove up to 10% of a donor colony so as to minimise harm[10]. In this way collection of transplant material should not adversely impact the reproductive output or survival of colonies on the source reef. A few tens of new colonies may be generated from the excised fragment(s) from each donor colony, thus potentially increasing the number of independently growing colonies by several times. A recent study suggests that using around 30–50 randomly sampled source colonies (or corals of opportunity) of a target species (widely spaced to avoid sampling clones) would allow you to retain a major proportion of the original genetic diversity of a population (see section 2.3)[11].

To make the best use of the original coral source material we recommend that once grown to an adequate size, a proportion of nursery-reared or directly transplanted corals are then used as sources of future fragments, taking care to maintain a diversity of genotypes. This will help minimise the need for damage to the natural reef and multiply up the potential benefits of the original removal of source material from the reef.

Finally there are projects where the decision has already been taken to severely impact or destroy an area of reef such that any corals not moved will probably die. In such mitigation projects where corals are being relocated, as much of the reef community (not just corals) as possible should be saved with priority given to species that do not rapidly colonise bare sites. For example, "fouling organisms" such as barnacles, sea squirts, sponges, bryozoans and algae tend to need little encouragement. Given that some transplanted corals will die during the relocation process, it may be sensible to fragment a proportion of the colonies being moved and either rear these fragments in *in-situ* nurseries or transplant directly to make up the expected losses from mortality.

Which species to transplant?

Careful selection of which coral species to transplant is one of the most crucial steps in successful restoration. When considering which species to use, the most logical choice is one that occurs naturally at the rehabilitation site and is relatively common on nearby potential source reefs. This does away with a lot of guesswork with respect to the ability of the species to survive in the ambient environmental conditions at the rehabilitation site. However, beware that transplants from the same species but collected from a different environment (i.e. different depth or wave exposure) might not be well-adapted to the transplant site (e.g. coral fragments collected from a reef slope population and transplanted to a back-reef environment might not survive well). Thus not only should you focus on species that are known to survive at your rehabilitation site but you should make sure that these are sourced from sites with environmental conditions that are as similar as possible to the rehabilitation site.

Species that are known to have occurred naturally at the rehabilitation site in the recent past may also be considered. There may be dead coral heads or piles of coral rubble at the site that indicate which species or genera were living at the site in the recent past. Coral species that have died out locally due to an episodic disturbance (e.g. bleaching event, storm, tsunami, or blast-fishing) are better candidates for reintroduction via transplantation than those that have died out due to chronic stresses (e.g. pollution, siltation). The altered environmental conditions will most likely prevent the latter set of species from recolonising the area, unless management measures have been taken to reduce the chronic stresses. Stresses could also be due to natural factors such as sand bars forming and reducing water circulation and the past-history of the site needs to be investigated during the project design phase (sections 2.2–2.3) so that risks to transplants can be assessed (Chapter 3).

In rare cases almost no coral may remain at the rehabilitation site and there be no local recollection of what types of coral were there before. In such cases you should seek a nearby undegraded site with similar environmental conditions to act as a "reference site" to guide your choice of species. If you have no idea whether coral previously occurred at your proposed "rehabilitation" site and are unable to find a nearby site in a similar environment that has a coral community, then you are taking an irresponsible, high-risk gamble if you attempt to transplant corals to your site. In such circumstance we recommend that you find another rehabilitation site. A reference site can also guide your choice of how far apart to place the transplants.

Fast-growing branching coral species, such as acroporids and pocilloporids, tend to be more susceptible to disturbances than slower growing massive and submassive species. In the Indo-Pacific they also tend to dominate recruitment and thus can often re-establish populations quickly once anthropogenic impacts are addressed by management. *Acropora* corals are known to bleach readily and are among the preferred prey of Crown-of-thorns starfish, *Drupella* snails and corallivorous fish. It is thus risky to undertake transplantation using predominantly such species, and better to use a cross-section of corals of different growth forms, growth rates and families that are adapted to your rehabilitation site in order to reduce risk (Table 3.1). Clearly, if you are trying to restore a community of *Acropora* thickets growing on sand, then you have little choice of species, but you should be aware of the inherent risks in what you are trying to achieve.

Message Board

- Coral transplantation is just one option available to coastal or MPA managers considering how to rehabilitate coral reefs.

- In most cases, *passive* restoration via good management of reef resources is the option of first choice.

- Coral transplantation should generally be seen as an approach of last resort.

- A crucial prerequisite for coral transplantation is that significant local anthropogenic impacts on the reef are under some form of effective management.

- A key issue for mitigation projects, where threatened corals are to be relocated to a new site, is finding a suitable site to which to move them. If corals are not naturally present at a site, then conditions for coral survival are unlikely to be conducive at that site.

- Artificial attachment of transplants should generally be regarded as an interim stabilisation measure to allow the corals time to grow tissue onto the substratum ("self-attach"). Once transplants have self-attached, the chance of detachment is greatly reduced.

- Fast-growing branching species may provide a rapid increase in coral cover and topographic complexity, but they tend to be more susceptible to bleaching, disease and coral predators.

- The primary aim of coral transplantation in reef rehabilitation should generally be to *assist* the natural recovery of a coral reef.

At what time of year should coral transplantation be carried out?

As discussed in Chapter 3, in order to minimise risks to transplants you should try to avoid transplanting immediately before the stormy season and at the time of year when sea temperatures are highest. The longer that transplants have to self-attach to the substratum before the period of worst weather (e.g. the tropical cyclone/hurricane/typhoon season) the better. Some acroporids can readily self-attach within a month and for slower growing species a majority of fragments or colonies appear to be able to self-attach within 3 months[12]. During the warmest months of the year bleaching is more likely and disease more prevalent and the additional stress of transplantation can lead to poor survival, thus it is a sensible precaution to try to avoid transplanting in the months just before and during the annual warm period (see Table 3.1). Trying to avoid both stormy periods and the warmest months may be difficult at some locations where calm periods coincide with warm sea temperatures. A compromise may be to carry out transplantation during the coolest months in sheltered reef environments (i.e. within protected lagoons, etc.) where corals will be protected from storm waves, and prior to the calmest months in more exposed environments where water flushing will hopefully prevent corals from becoming heat stressed soon after transplantation.

An Acropora colony about one month after being attached with cement, showing the rapid self-attachment (bluish growth) onto the substratum around the base (Goro Nickel).

Role of artificial substrates

Artificial substrates (often so-called "artificial reefs") are often used in scientific experiments involving coral transplantation in order to have standard units of "reef" that can be validly compared. However, there is usually little justification for using them in coral reef restoration, given the underlying assumption that you are trying to rehabilitate a natural ecosystem, not replace it. In some special cases, usually in relation to tourism, artificial structures may be used to create additional attractions for SCUBA divers or to divert novice divers, who may carelessly damage fragile corals, away from the natural reef.

6.4 Transplantation methods for different environments

The method of transplantation that will be most effective in a given case will depend on: 1) the species of coral being transplanted, 2) the nature of the substrate at the transplantation site, and 3) the environment at that site. In this manual we are primarily concerned with restoring natural reefs but methods of transplantation to these should usually apply equally to artificial structures. For a method to be effective, it must ensure good transplant survival with both *in-situ* mortality and losses due to detachment being minimized. Key factors determining overall survival are the care taken to minimize stress to the corals during transplantation (see earlier chapters and *Reef Restoration Concepts and Guidelines*[5]) and that taken to ensure that transplants are securely attached at the rehabilitation site. The difficulty of achieving secure attachment depends in turn on the environment, becoming progressively more difficult as sites become increasingly exposed. On the one hand, in calm lagoonal environments it may be sufficient to wedge (e.g. Ch. 8: Case study 2 and 8) or tie in place with polythene string[13]; on the other, in reef crest sites exposed to the waves one may be doomed to lose a considerable portion of transplants even when fixed with epoxy or cement. It is not uncommon for 50% of transplants to be lost from shallow exposed sites. Most corals require a consolidated hard substrate on which to grow but a few species can survive on rubble (e.g. fungiids) or sand (some acroporids) in relatively calm environments (e.g. Ch. 8: Case study 2).

Artificial attachment of transplants by whatever method should generally be regarded as an interim stabilisation measure to give the coral time to securely self-attach by growing tissue over the substratum. So whilst the attachment should be made as robust and long-lasting as possible, every effort should be made to encourage the transplant to grow onto the substratum (natural or artificial). If the coral does not self-attach, most man-made attachments will fail within a few years. Generally, living coral tissue in contact with relatively clean hard substrate will grow onto it and so a few points of tissue contact are recommended to promote self-attachment. Sponges and other fouling organisms that will hinder self-attachment (e.g. Ch. 8: Case study 6) and potentially overgrow the coral transplant should be removed from the immediate vicinity of attachment points. Only living coral tissue can grow onto the substrate so care must be taken to make sure that bases of transplants are trimmed of dead patches or significant areas of exposed old skeleton. These can both be targeted by bio-eroders and hinder self-attachment of the coral base. (Note that exposed skeleton can be coated in epoxy to prevent attack by borers.)

The coral transplant will not be able to self-attach if it is subject to any movement (e.g. rocking from side to side as waves pass over). Thus it is crucial that transplants are securely held in place, with no movement, until they have had time to self-attach. Similarly, where wires, monofilament fishing line or cable-ties are used to attach coral fragments, these can be effective, with coral tissue growing over them and onto the substrate, but only when there is no movement. If the fragments are moved by wave action then the wires, lines or cable-ties can abrade the coral tissue and either the fragments work loose or the attachments eventually break.

The most expensive part of *active* reef restoration is normally the act of transplantation (see Chapter 7). This is largely due to the time taken to attach the transplants securely and the likely need for boats and scuba diving. Similarly, where nursery rearing of coral fragments is undertaken (Chapter 4), the second most costly stage is attaching the coral fragments to substrates in the nursery. To improve the cost-effectiveness of these two stages, there has been considerable research into rearing fragments on substrates such as commercially available plastic wall-plugs and tubing (for branching corals) and plastic mesh or segments of PVC pipe (for massive, submassive, encrusting corals) that can be readily deployed on degraded reefs (section 4.4). Early results (Ch. 8: Case study 5) suggest that coral colonies grown on wall-plugs, which can then be slotted into pre-drilled holes in the reef, can be deployed about five times as fast as colonies or fragments being attached using epoxy putty.

In the next section we consider methods of transplantation in:

1. Sheltered environments where corals can be left unattached or just loosely attached,

2. Relatively sheltered environments where corals can be wedged in holes and crevices without the need for adhesives or cement,

3. Areas where corals need to be securely attached with cement or adhesives, and

4. We discuss the use of wire, monofilament line and cable-ties to secure coral fragments.

A Porites cylindrica fragment newly wedged into a hole on a degraded reef at a sheltered lagoon site during a community restoration project (D. dela Cruz)).

1. Unattached and loosely attached transplants in sheltered environments

Fungiids (mushroom corals), several species of *Acropora* (including *A. palmata* and *A. cervicornis* in the Caribbean) and *Porites*, some species of *Montipora* and eastern Pacific *Pocillopora damicornis* have been reported to reproduce successfully by fragmentation[9], typically as a result of storm damage or by the actions of other reef organisms. (Note that fungiids live naturally unattached and their fragments do not need to be attached to the substrate, as was reportedly done by one enthusiastic group of reef restorers!) The faster growing species can reattach to the reef substrate and can heal fragmentation wounds within a few weeks and as a result have been popular candidates for transplantation. In some low-energy environments such as back-reef zones, lagoons and protected embayments, fragments of such species can survive and grow without being artificially attached to the substrate. In these limited situations, coral transplantation may be effected by the scattering of fragments of an appropriate species on the substratum. Some staghorn corals (*Acropora* spp.), such as *A. intermedia* and *A. muricata*, which naturally occur on sand, can be transplanted by simply burying the base in the sand, provided the branch is long enough (about 50 cm). However, even in relatively sheltered environments stabilization by linking coral branches by polythene string (so

that they are held in place by the collective weight of all attached branches) has been shown to increase survival and growth[13] and one of the lessons from Case study 2 (Chapter 8) was that transplants should have been placed closer together in small clonal patches so that individual branches could give each other mutual structural support (and eventually fuse). Soft sediments also offer the risk of abrasion and burial, sometimes as a result of the activities of burrowing animals.

2. Attachment without adhesives or cement in relatively sheltered environments

In relatively sheltered environments where there are natural holes or crevices in the reef, smaller (< 10 cm) coral branches or fragments can be slotted into them (a biodegradable stick such as a sliver of bamboo can be used as a wedge to make the attachment more secure). Care should be exercised to maximize the direct contact of the living portion of the transplant to the calcium carbonate rock substrate in order to hasten natural self-attachment by the coral. Scraping the substrate around prospective attachment points clean of algae or encrusting sessile invertebrates (e.g. sponges or tunicates) with a small wire brush can assist self-attachment. Holes can also be created artificially using a hand-held auger or hand drill or compressed-air drill if the reef substrate is very hard. Larger fragments will generally require adhesives and even initial support wires for stability while waiting for self-attachment to occur. Such areas with natural holes at sheltered shallow sites are particularly suitable for community restoration projects[14].

3. Attachment with Portland cement and epoxy adhesives

The use of adhesives and cement for fixing corals directly onto hard substrate is probably the most common method. The technique is labour-intensive and analysis of a range of studies suggest that only about 5–10 fragments or colonies can be attached per person-hour once all necessary peripheral activities are taken into account. A range of adhesives have been used in transplantation experiments and restoration activities (Table 6.1). The choice of adhesive depends on local availability, the environmental conditions at the restoration site, the size and morphology of the corals, the amount of coral that needs to be attached, and the manpower and financial resources available to undertake the restoration.

Squeezing cement and sand concrete mix (with plaster of Paris added to speed setting) from a plastic bag onto a degraded reef prior to attaching a transplant (R. Dizon). Robust plastic pastry bags can also be used to pipe cement onto substrate.

Cement, usually in the form of sand-cement concrete, is considerably cheaper than epoxy and thus may be the material of choice for the attachment of large massive and submassive corals and for repairing reef framework damaged by ship-groundings or tsunamis. Larger colonies and fragments with a relatively high basal area to volume ratio are likely to stay in place long enough for the concrete to set. Smaller transplants or those with a narrow base may need to be embedded in the concrete or temporarily held in place with wire tied to nails at higher energy sites (e.g. Ch. 8: Case study 6) to ensure they do not get moved by wave action prior to the concrete setting. Admixtures can be used to speed up the setting process.

Coral tissue directly in contact with unset concrete is likely to be killed and care should be taken to minimize contact between live coral tissue and the concrete and not to spill any concrete on the coral. Concrete can be premixed on land or on a boat in batches sufficient for one dive's worth of coral transplants and then placed in strong plastic bags for transport underwater. An appropriate amount can then be squeezed onto each attachment site as each transplant is fixed in place. In general the amount of concrete should just be sufficient to ensure that each transplant is secure and make it as easy as possible for basal tissue from the transplant to grow outwards onto the substrate. To ensure good adhesion to the substrate it may be advisable to scrub the point of attachment clean with a wire brush prior to applying the concrete mix.

Diver preparing to place a whole Acropora coral colony onto a pile of marine cement at a transplantation site during the translocation of 2000 coral colonies rescued from the Goro Nickel harbour development in New Caledonia (Goro Nickel).

Concrete has also been used to anchor large transplanted *Acropora* branches in sand during attempts to restore sand-based *Acropora intermedia* or *A. muricata* thickets by digging a small depression and squeezing the premixed concrete from a plastic bag directly into the depression to minimize its dispersion. Some sand can be poured back over the concrete as the branch is embedded a few centimetres into the cement. This avoids the use of unsightly concrete blocks to stabilize such transplants. In such cases, setting up small clonal patches which allow adjacent branches to fuse and give each other mutual support when they come into contact will help with long term stability.

Ordinary Portland cement mixed with sand and freshwater (try to avoid using saltwater as this may interfere with the setting process and strength of the concrete) has been widely used, sometimes with admixtures to alter the rate of setting of the concrete. Type II Portland cement or specialist sulphate resistant marine cements with microsilica-based additives are recommended for use in the marine environment and can be used if available locally.

While a number of adhesives have been employed to attach coral fragments to substrates underwater, various brands of two-part epoxy putty (e.g. AquaMend®, Epoxyclay Aqua™) appear the most suitable in terms of ease of use and cost-effectiveness[15]. Epoxy putty comes in small sticks (e.g. 60–70 g) and the chemical reaction between the two parts does not begin until you mix the two parts together. It has the advantage that for each transplant you only break off the amount you need. You knead the two parts together with your fingers until thoroughly mixed and then place the putty on the attachment surface (preferably pre-cleaned with a wire brush) and press the coral gently into the epoxy putty, using the putty to cover any exposed coral skeleton at the base. Once mixed the epoxy remains workable for several minutes (the time depending on the type used) and is usually set within 10–30 minutes. Thus epoxies with different setting times can be used according to your needs. As with cement, you should try to use the minimum amount of adhesive necessary to fix the transplant securely to the substrate and make it as easy as possible for the coral's basal tissue to grow outwards onto the substrate. Care should be exercised to minimise direct contact between live coral tissue and the epoxy. Some brands appear more toxic to corals than others when not set. You want a brand that is designed for underwater use, sets reasonably fast, is easy to mix underwater, does not stick to your fingers too much, and is stiff enough to support your transplants once in place[16].

Where lots of corals are being transplanted, the cost of epoxy putty can be substantial. Cheaper marine epoxy of the kind that is bought in two cans and then mixed before use attaches corals well but is difficult to handle underwater and ultimately more time consuming and thus costly to use.

Table 6.1 Adhesives and other materials used in attaching coral fragments to the reef substrate.

Adhesive/ attachment material	Advantages	Disadvantages	Notes for restoration
Portland cement, sand-cement mixes (often with admixtures (e.g. Sikacrete®W) to alter setting times, enhance cohesion, improve flow properties, reduce wash-out and improve strength)	• Cheap and widely available • Provides secure attachment • Can be mixed on land, placed in plastic bags or pastry bags with nozzles and applied underwater	• Tendency to use excessive cement to secure transplants (with result that coral takes a long time to self-attach to the natural substrate)	• Recommended for large fragments or whole colonies • Adding admixtures or Plaster of Paris to the mix can shorten the setting time • Can secure fragments initially with wire while cement cures (24 hours)
Epoxy putty (e.g. AquaMend®, Epoxyclay Aqua™)	• Provides secure attachment • Can cure in around 10–15 minutes (time depends on type) • Easy to mix and apply underwater • Suitable for attaching transplants *in situ* underwater • Can be used to cover exposed skeleton of transplant to prevent fouling by algae or attack by bio-eroders	• Relatively costly (compared to cement) • Only small amounts can be mixed and used at a time due to its quick setting time	• Try to ensure there are several points of contact where coral transplant can self-attach to the natural substrate
Marine epoxy (two part - resin and hardener)	• Provides secure attachment • Cheap and widely available	• Cannot be prepared underwater • Tends to have a longer setting and curing time than epoxy putty (around 30 minutes) • Difficult to apply (tendency to use more adhesive per coral than necessary)	• Try to ensure there are several points of contact where coral transplant can self-attach to the natural substrate
Cyanoacrylate glue (e.g. Super Glue®, Loctite®, Reef Glue)	• Widely available • Cures within seconds • Only a small amount required per small fragment • Good for attaching small fragments to substrates for nursery rearing	• Difficult to apply effectively underwater (even using a gel type) • Fragments prone to dislodgement in high energy areas • Not generally appropriate for large fragments	• Commonly used to attach small fragments to artificial substrates (out of water) for nursery rearing • Better used for species which self-attach quickly
Wire (stainless steel, insulated single strand copper wire, Monel 400, etc.)	• Cheap and widely available. If coral is held securely, it can grow over wire within weeks or months	• Causes breakage if lashed too tightly • May be difficult to attach to natural substrate	• Best used as a temporary measure that allows self-attachment
Monofilament fishing line	• Cheap and widely available • If coral is held securely, it can grow over line within weeks or months	• Adds to unnatural material in the reef • Tying on transplants can be labour-intensive	
Cable-ties	• Easy and quick to use and widely available • If secure, coral grows over straps within weeks or months	• Adds to unnatural material in the reef. • Tendency to loosen in areas exposed to wave action	• Advisable to have several lengths of cable ties on hand

4. Use of wire, fishing line and cable-ties

Insulated electrical wire, stainless steel wire, monofilament fishing line and cable-ties have all been used to attach coral transplants, usually either to artificial structures (e.g. Ch. 8: Case study 3) or to artificial substrates for nursery rearing (e.g. 20 cm x 5 cm x 1.5 cm limestone slabs[17]), which are later transferred to the reef once corals have grown. They may also be suitable for attaching fragments to thick dead coral branches or to nails or metal stakes fixed in the reef. However, dead coral branches are likely to be subject of bio-erosion and thus prone to eventual collapse, whilst introducing nails and stakes appears an aesthetically poor option where less obtrusive methods of attachment can be used. However, if the corals are securely fixed and not subject to movement by waves, then they generally will quickly overgrow the wire, line or cable-tie and in higher energy environments tying transplants to nails in the substrate may be a useful option. If there is any movement of corals then the material is likely to abrade the coral tissue and fragments may work loose. Thus care needs to be taken when tying corals in place with these methods. Unused loose ends of cable-tie, wire or fishing line should be cut off and removed from the reef.

Community members attaching coral fragments found lying detached on the seabed to a "rescue" station with cable-ties. These will be used as sources of transplants in future community restoration activities (M.V. Baria).

Good Practice Checklist

✓ Ensure that corals being transplanted to a rehabilitation site are well-adapted to survive at that site (appropriate species and from a similar environmental setting).

✓ Make use of "reference sites" to inform the selection of species and provide estimates of the density of transplants that may be appropriate.

✓ Attempt to transplant a mix of common species, growth forms and families adapted to your rehabilitation site to increase the resilience of your transplant community.

✓ Transplant corals at times of year when they are likely to be least stressed and prone to disease (i.e. outside the warmest months and not during the main spawning time for seasonal spawners – see Figure 5.2).

✓ For more exposed rehabilitation sites, avoid transplanting in the months just before or during the stormy season.

✓ Minimise exposure of coral transplants to air, sun or water temperatures above ambient; keep species and genotypes separated during transport; minimise handling; maintain good water quality during transportation.

✓ Try to encourage self-attachment of transplants to the reef substrate by transplanting them such that living coral tissue is in contact with the substrate. Try to minimise the amount of epoxy putty or cement between the transplant and the reef substrate.

✓ Carry out routine monitoring of your transplants and maintenance visits. These are likely to be highly cost-effective given the expense of carrying out transplantation and could prevent wholesale loss of transplants to predators.

✓ If feasible, set up and monitor a few comparable "control" areas where no active restoration has been attempted. These provide a clear baseline against which you can evaluate the cost-effectiveness of your coral transplantation.

6.5 Nursery rearing of corals on substrates for transplantation

Rearing of coral fragments in nurseries (Chapter 4) prior to transplantation makes much better use of a given amount of coral source material and provides an opportunity to establish the transplants on substrates that can be readily attached at a degraded reef site. This is easiest for branching species that can be grown in plastic wall-plugs or pieces of plastic hose pipe (or any plastic tubing), which are later used for attachment to the reef (section 4.4). Using a hand auger, hand-drill or compressed-air drill depending on the hardness of the reef rock, holes of a diameter and depth such the substrate will fit snugly can be drilled in the reef and the small colonies slotted into them. If the fit is not exact then a little epoxy putty can ensure secure attachment. The advantage of using wall-plugs is that the correct drill bit sizes are given for each size of wall-plug and the living tissue at the base of the transplant will generally end up in contact with the reef substrate around the top of the hole, promoting self-attachment.

For massive, foliose or encrusting corals grown in nurseries on plastic mesh or pieces of PVC pipe (section 4.4), these substrates can be attached with either epoxy putty, flat headed masonry nails or large staples. For mesh the masonry nails or staples can be inserted through the interstices in the mesh whereas the PVC segments have pre-drilled holes at their corners through which they are tied to the nursery rearing trays. At present the long-term success of transplanting using mesh and PVC pipe substrates is still under investigation. However, as long as the corals are able to self-attach the method appears to have considerable potential.

Deploying transplants reared in rope nurseries

Large numbers of coral fragments can be reared cheaply in rope nurseries (Chapter 4). Given that attachment of individual transplants at the rehabilitation site is the most expensive part of active restoration, being able to transplant the grown colonies *en masse* to a degraded reef whilst still attached to the rearing ropes appears to offer considerable potential as a cost-effective method of deploying transplants. The advantage is that a rope with several tens of corals can be rapidly laid over the coral rock substrate and then fixed securely to it with galvanized masonry nails. If this is done carefully each coral colony should be pressed against bare substrate and will be able to self-attach. Some species will do this readily, others less so. Once colonies

---

Box 6.1 Tools needed

1. A hammer and chisel are often needed to break off fragments of massive corals. For this purpose, an ordinary light hammer and a cold chisel will do.

2. Side-cutting pliers are useful for excising branches and trimming fragments.

3. A hand-auger can be used to bore holes in soft calcium carbonate rock either for attaching small coral colonies reared on substrates such as wall-plugs or for slotting branching coral fragments directly into the reef. You can make one from a large Allen wrench (10–15 mm diameter) with a plastic handle by grinding

A stick of two-part epoxy putty and hand-auger made by welding a steel handle to a cold chisel and covering the handle with plastic tube (P. Cabaitan).

two sides of the tip to make the end like a chisel. Another way is to use a cold chisel and weld handles to the top end which can then be covered by plastic or rubber tubing. The resulting "T"-shaped tool allows good purchase. Still another way is to weld handles to the head of a very large screw which can then be used as an auger.

4. A compressed air driven underwater drill may be needed for drilling holes in hard calcium carbonate rock, or for making many holes in softer rock. One large enough to hold a 10 mm drill bit should be purchased. Several brands are available from specialist hardware stores, but these may not be readily available and need to be ordered so advance planning may be required. While these drills are not prohibitively priced, they tend to have a limited life span because of their use underwater. To prolong their usefulness, they must be thoroughly rinsed in freshwater <u>immediately</u> after use (do not allow the drill to dry with seawater on it), then quickly lubricated with an appropriate oil. You can immerse the entire drill in kerosene (or vegetable oil as a less toxic alternative) until the next use or you can dismantle the drill and thoroughly spray all working parts with a lubricant (e.g. WD40™).

Care of tools
Seawater being corrosive for most metal tools, these must be thoroughly rinsed in freshwater after each day's use. This will prolong the useful life of most tools. Where available, stainless steel tools may be used to reduce rusting.

have self-attached, exposed rope could be removed. To promote self-attachment the reef substrate ought to be scraped clean of fouling organisms beneath each colony (trying to minimise damage to existing sessile invertebrates) and care must be taken not to place colonies on top of any existing corals or on top of other sessile fauna. This may be easier said than done because the spacing of corals on the rope is pre-determined and it may be difficult to avoid doing some collateral damage to the existing fauna when deploying rope-reared corals. At present, we know that corals can be deployed in large numbers this way but rates of self-attachment, survival and collateral damage remain unquantified.

A Pocillopora damicornis colony being grown in a rope nursery from a small fragment (J.R. Guest). Research is still needed on the best way of deploying corals reared in such nurseries to rehabilitation sites.

6.6 Monitoring and maintenance of transplants

Having invested considerable time and resources in establishing transplants at a rehabilitation site (probably US$ 10,000s per hectare transplanted – see Chapters 7 and 8), it makes sense to monitor the transplants and carry out maintenance if needed. As discussed in Chapters 2 and 3, monitoring is needed for adaptive management as well as for assessing progress towards the goals of the rehabilitation project and giving feedback to stakeholders and the local community or client. Regular visual checks on the status of the transplants are enough to identify problems (e.g. *Drupella* or Crown-of-thorns starfish attack) that may need adaptive management (e.g. maintenance action to remove the predators) whereas semi-annual or annual systematic surveys may be needed to show progress towards goals (such as increasing coral cover or build up of reef fish biomass). To show that any changes are due to the transplantation rather than other factors, it is generally desirable to monitor a few similar areas nearby which have not been subject to active intervention.

The parameters that you should measure during monitoring will depend on the aims and objectives of your project. The most common parameters are:

Living coral cover: Demonstrating an increase in living coral cover may be considered as a basic criterion for success. Various methods have been devised for measuring coral cover[18-19] and it is important to use the same method for the baseline (pre-transplantation) and subsequent monitoring surveys to ensure comparability.

Reef fish diversity and abundance: For many people, the changes in the fish communities are of greater economic and touristic interest than the recovery of coral cover, although there is usually a positive correlation between the two. The types of fish and detail of monitoring will depend on project aims and methods should be selected from standard manuals[18-19] to suit your aims. For example, if an increase in reef fish biomass is part of the project aims then you will need to estimate lengths as well as numbers of fish of each species per unit area and use FishBase length-weight relationships (www.fishbase.org) to calculate biomass.

Environmental measurements: It may be useful to monitor certain environmental variables such as temperature at the transplant site to establish the typical annual temperature regime for the site and give warning of unusually high temperatures (see also Table 3.1). In a warming event, there may be little you can do except perhaps shade your transplants (e.g. by floating plastic mesh on the sea surface above them) but at least you will know the cause of coral transplant mortality (for example if corals die from warming induced bleaching).

If your project has social and economic objectives then there may also be a need for surveys to assess whether economic and social objectives are being met.

Maintaining transplantation sites

At sites that are overfished, predation of transplants by fish is seldom a problem, but at sites where fish are abundant, new transplants may suffer considerable damage from parrotfish, some larger wrasse species (e.g. *Coris*, larger *Bodianus*, *Novaculichthys*), triggerfish and butterflyfish. Fish appear to be attracted to freshly attached coral transplants with some species feeding directly on the coral polyps and others seeking invertebrates such as molluscs embedded in the coral skeleton. For this reason some workers have found it necessary to protect transplants with plastic mesh cages or netting for several days after attachment. You can carry out a small trial transplantation at your site to test whether fish attack is likely to be a problem and caging will be needed. This will clearly add to the costs of the transplantation. As long as transplants are not detached by the fish they may recover from this initial grazing (Ch. 8: Case study 5).

Generally, some maintenance of transplant sites is desirable, if not absolutely necessary for ensuring the success of your restoration efforts. The need for maintenance activities will vary from site to site depending on the state of the local environment. If water quality is good and fishing pressure moderate then little maintenance may be needed as control of macroalgae and coral predators will essentially be provided as an ecosystem service. On the other hand if water quality is poor and fishing pressure high, considerable maintenance may be needed. Indeed, in such circumstances your transplantation may be a high risk venture that is unlikely to be sustainable.

The following maintenance activities should be considered:

- Reattachment of detached transplants. Depending on the method of transplantation used and the amount of care taken, some corals may become detached as a result of physical disturbance (e.g. waves, fish, divers).
- Removal of loose fouling materials, whether in the form of man-made flotsam (e.g. garbage, fishing net) or natural items like loose seaweed fronds.
- Removal of coral predators such as some gastropods (e.g. *Drupella*, *Coralliophila*) and some echinoderms (e.g. *Acanthaster*, *Culcita*).
- Removal of fouling organisms, notably fleshy or filamentous macroalgae, sponges and tunicates that may overgrow your transplants.

Managing transplants over time

To minimise the original damage to donor colony reefs and maximise the multiplicative effect of transplantation, successfully transplanted corals can be used as future donor colonies for further transplantation efforts in additional areas. As with the original donor colonies, care must be taken not to dislodge the transplanted colonies and to excise no more than about 10% of the new donor colonies.

It may be necessary to periodically add transplants to increase the original number deployed. This may happen if the original number available for transplantation was limited or if mortality has been higher than anticipated. In the latter case, you should carefully consider whether the causes of the mortality have abated before putting more corals at risk. Alternatively, there may be a desire to increase the number of species transplanted at the site, thus increasing the diversity of the restored community. Initially you might start by transplanting the hardiest species and, if these thrive, then add less hardy species.

References

1. Adger, W.N., Hughes, T.P., Folke, C., Carpenter, S.R. and Rockström, J. (2005) Social-ecological resilience to coastal disasters. *Science*, 309, 1036-1039.
[Download at: www.sciencemag.org/cgi/reprint/309/5737/1036.pdf]

2. Edwards, A.J. and Clark, S. (1998) Coral transplantation: a useful management tool or misguided meddling? *Marine Pollution Bulletin*, 37, 474-487.

3. Rinkevich, B. (2008) Management of coral reefs: We have gone wrong when neglecting active reef restoration. *Marine Pollution Bulletin* 56, 1821-1824.

4. Maragos, J.E. (1974) Coral transplantation: a method to create, preserve and manage coral reefs. Sea Grant Advisory Report UINHI-SEAGRANT-AR-74-03, CORMAR-14, 30 pp.
[Download at: nsgl.gso.uri.edu/hawau/hawaut74005.pdf]

5. Edwards, A.J. and Gomez, E.D. (2007). *Reef Restoration Concepts and Guidelines: making sensible management choices in the face of uncertainty*. Coral Reef Targeted Research & Capacity Building for Management Programme: St Lucia, Australia. iv + 38 pp. ISBN 978-1-921317-00-2.
[Download at: www.gefcoral.org/Portals/25/workgroups/rr_guidelines/rrg_fullguide.pdf]

6. Kilbane, D., Graham, B., Mulcahy, R., Onder, A. and Pratt, M. (2009) Coral relocation for impact mitigation in Northern Qatar. *Proceedings of the 11th International Coral Reef Symposium*, 1248-1252.
[Download at: www.nova.edu/ncri/11icrs/proceedings/files/m24-12.pdf]

7. Seguin, F., Le Brun O., Hirst, R., Al-Thary, I. and Dutrieux, E. (2009) Large coral transplantation in Bal Haf (Yemen): an opportunity to save corals during the construction of a Liquefied Natural Gas plant using innovative techniques. *Proceedings of the 11th International Coral Reef Symposium*, 1267-1270.
[Download at: www.nova.edu/ncri/11icrs/proceedings/files/m24-17.pdf]

8. Van Treeck, P. and Schuhmacher, H. (1998) Mass diving tourism – a new dimension calls for new management approaches. *Marine Pollution Bulletin*, 37, 499-504.

9. Highsmith, R.C. (1982) Reproduction by fragmentation in corals. *Marine Ecology Progress Series*, 7, 207-226.
[Download at: www.int-res.com/articles/meps/7/m007p207.pdf]

10. Epstein, N., Bak, R.P.M. and Rinkevich, B. (2001) Strategies for gardening denuded coral reef areas: the applicability of using different types of coral material for reef restoration. *Restoration Ecology*, 9 (4), 432-442.

11. Shearer, T.L., Porto, I. and Zubillaga, A.L. (2009) Restoration of coral populations in light of genetic diversity estimates. *Coral Reefs*, 28, 727-733.

12. Guest, J.R., Dizon, R.M., Edwards, A.J., Franco, C. and Gomez, E.D. (2009) How quickly do fragments of coral 'self-attach' after transplantation? *Restoration Ecology*, 17 (3) [doi: 10.1111/j.1526-100X.2009.00562.x]

13. Lindahl, U. (2003) Coral reef rehabilitation through transplantation of staghorn corals: effects of artificial stabilization and mechanical damages. *Coral Reefs*, 22, 217-223.

14. Gomez, E.D. (2009) *Community-based restoration: the Bolinao experience*. Coral Reef Targeted Research & Capacity Building for Management Program, St Lucia, Australia. 4 pp.

[Download at: www.gefcoral.org/LinkClick.aspx?
fileticket=5ysYNhGL45I%3d&tabid=3260&language=
en-US]

15. Dizon, R.M., Edwards, A.J. and Gomez, E.D. (2008)
Comparison of three types of adhesives in attaching coral
transplants to clam shell substrates. *Aquatic Conservation:
Marine and Freshwater Ecosystems*, 18, 1140-1148.

16. Reef Ball Foundation (2008) *A step-by-step guide for
grassroots efforts to Reef Rehabilitation.* Reef Ball
Foundation, Inc., Athens, Georgia, USA. 133 pp.
[Download at: www.reefball.org/stepbystepguidetoreef
rehabilitation/DraftGiude.pdf]

17. Heeger, T. and Sotto, F. (eds). (2000) *Coral Farming: A
Tool for Reef Rehabilitation and Community Ecotourism.*
German Ministry of Environment (BMU), German Technical
Cooperation and Tropical Ecology program (GTZ-TÖB),
Philippines. 94 pp.

18. Hill, J. and Wilkinson, C. (2004) *Methods for Ecological
Monitoring of Coral Reefs: A Resource for Managers.
Version 1.* Australian Institute of Marine Science, Townsville,
Australia. vi +117 pp. ISBN 0 642 322 376.
[Download at: data.aims.gov.au/extpubs/attachment
Download?docID=1563]

19. English, S., Wilkinson, C. and Baker, V. (1997) *Survey
Manual for Tropical Marine Resources.* 2nd Edition. Australian
Institute of Marine Science, Townsville, Australia. 390 pp.

Chapter 7.

Evaluating costs of restoration

Alasdair Edwards, James Guest, Buki Rinkevich,
Makoto Omori, Kenji Iwao, Gideon Levy, Lee Shaish

7.1 Introduction

"The costs involved in reef restoration projects are rarely fully assessed and reported. Few sources of information exist and even those that do exist do not generally identify all the relevant costs. There is therefore a need for a comprehensive costing framework that can be applied to future reef restoration schemes. This should give a detailed breakdown of all cost components in a consistent manner."[1] This chapter seeks to provide such a framework and show how it can be used both to assist planning of rehabilitation projects and to prioritise where research into reef restoration methodologies should focus.

Trying to discover how much reef restoration activities really cost is much more difficult than you might expect. Unfortunately, when asked about costs, many of those involved in restoration seem to think that the main objective is to show that their method is more "cost-effective" (by which they generally mean "cheaper") than other people's. Thus the time of some people involved in the project is not costed because they are "scientists" or volunteers, necessary equipment is not costed because it was "borrowed", SCUBA gear is not costed because it already belonged to a project participant, etc. This is not helpful to those planning to carry out reef restoration projects, who need to be able to estimate the real costs of what they are planning to do. People's time is not generally free (certainly not on a sustainable basis) and necessary items of equipment may not be borrowable at many locations. It is the recipient of the costing information who can make locally appropriate assumptions in terms of volunteer labour, free access to equipment, etc., not the supplier of the information.

There are three main reasons for carefully costing restoration projects. The first is so that others intending to carry out restoration can use the itemised costings to make realistic estimates of how much <u>their</u> restoration project may cost and judge what equipment, consumables and logistics may be required to achieve their project goals. The second is so that the cost-effectiveness of different techniques can be compared validly. The third is to identify the stages within a rehabilitation project which are responsible for most of the costs so that restoration research can focus on reducing those costs. In a coastal management context, the cost-effectiveness information can be expanded into a broader benefit-cost analysis (BCA)[2] to see whether *active* restoration is an efficient allocation of resources or whether the same funds might be better used, for example, to improve enforcement of existing (*passive*) management measures that promote reef resilience.

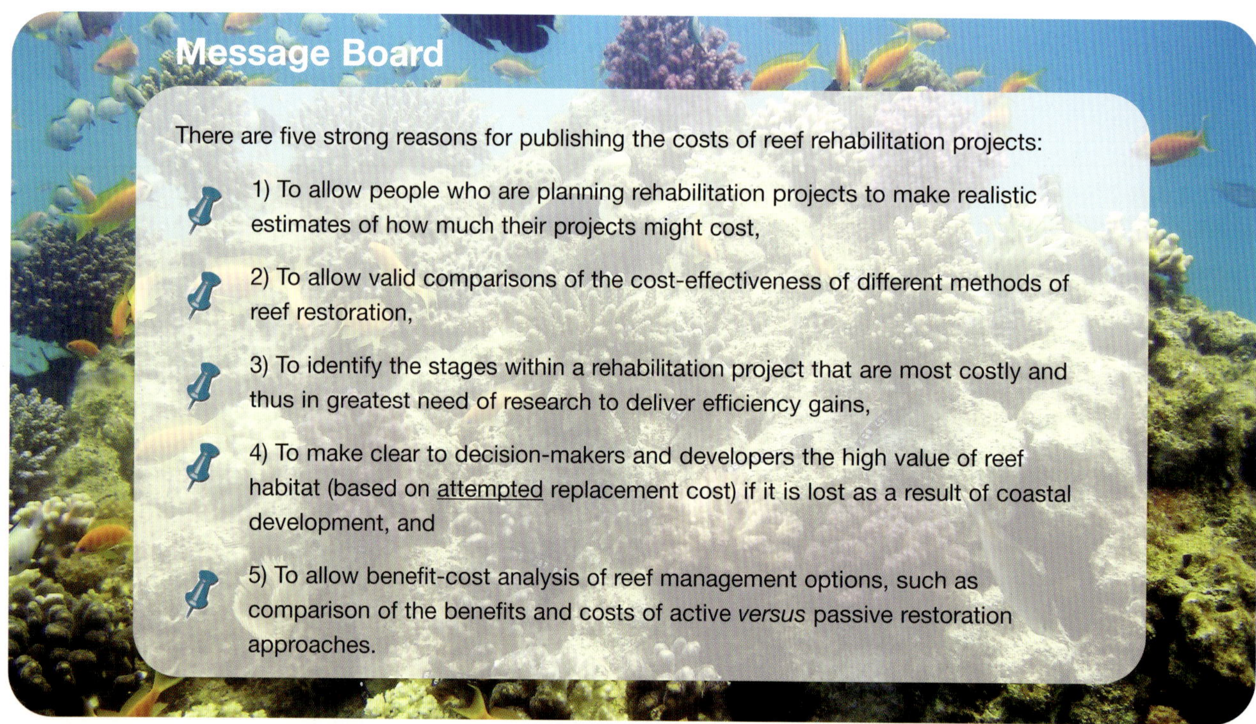

7.2 A framework for costing reef rehabilitation

In order to allow informed planning of reef rehabilitation projects, it is necessary to have estimates of personnel, equipment and consumables needed to implement and monitor a particular type of project. The nature of reef rehabilitation means that, in most projects, boats and SCUBA diving are likely to be involved. Both can entail considerable expenditure; thus in the example costings we present below, the numbers of boat-days and numbers of tanks of air for diving are central to estimating the required inputs. To allow a monetary cost per transplant to be calculated and compared, we have used various assumptions on local wage rates and expressed the results in US dollars to standardise.

Scaling up costs based on specific times required to perform particular tasks underwater can be misleading because all the peripheral activities (e.g. preparation of equipment, loading of boats, travel time to and from sites, etc.) tend to be overlooked. As an example, in one case-study featured in Chapter 8, the total time actually spent on a set of tasks was ~170 person-days (based on a 5-day week). However, using the stated times to undertake the activities (preparation of colonies for transplantation at 10 per hour assuming 6-hour working days, transplantation at 30 per hour assuming 4-hours diving per day), the tasks should have taken no more than 25 person-days. Even if one allows that half of the time spent was related to scientific documentation of the project, the tasks still took over 3 times as long as predicted from the deconstructed rates for individual focused activities. A manager who planned a restoration project based on the highly optimistic estimate of 25 person-days for these tasks would be justifiably upset if they actually took 85 person-days or longer. Underestimating costs benefits nobody and ultimately is likely to jeopardise projects by promoting adoption of unrealistic budgets.

Costing effort

Wage rates differ dramatically between countries so actual costs of employing people to do certain tasks are not readily transferable between projects. However, the time (person-hours) taken to do specific tasks is likely to be approximately the same from place to place. If you know how long it will take to carry out a task, then you can cost it based on local wage rates (or availability of volunteer labour). To allow comparisons between projects you need to standardise how you calculate time inputs. Table 7.1 provides one way of doing this so that, for example, one project's person-month is equivalent to another's. For example, just because one person on your project works 12 hours a day and seven days a week (i.e. 84 hours a week), does not mean that his/her working-week output can be emulated by everyone else, thus using that person as the basis for a person-week will be misleading to others planning projects; hence the need for some form of standardisation.

For different tasks, people with different skill levels may be needed and in some examples presented in this chapter we have divided person-hours between three skill levels. When planning a rehabilitation project the local wage rate for each skill level can be inserted to derive estimates of local personnel costs.

Table 7.1 Conversion table for standardising time inputs.

Person-years	Person-months	Person-weeks	Person-days	Person-hours
1	11	44	220	1760
	1	4	20	160
		1	5	40
			1	8

Separating set-up costs and operational (running) costs

Operational or running costs are perhaps more important from the point of view of the sustainability of restoration associated activities than set-up costs, because the latter can often be financed from outside sources as one-off donations, whereas ongoing funding is harder to obtain. Thus set-up costs need to be separated from running costs. Also, equipment and facilities created as part of setting up a project may have a life of three to five years or more, if maintained properly. Thus their costs may need to be treated separately when evaluating cost-effectiveness (e.g. spread or pro-rated over several years). The costs of materials needed to construct a coral nursery would clearly be classed as set-up costs.

If you will need to use SCUBA to collect corals or build and maintain a coral nursery, then you need to make sure that this is understood by stakeholders. You can either cost in the equipment needed for a SCUBA set up (compressor, tanks, etc.) and training (if required), or the estimated cost of necessary tank hire, air, etc. if purchased at local market rates (assuming, of course, that hiring is an option at your location). Since hire costs are likely to vary greatly from place to place, we suggest estimating needs in terms of numbers of air tanks, etc. needed for identified tasks involved in nursery set-up or maintenance, or in transplantation. The critical thing is to identify all necessary tasks and what equipment and consumables (and level of training) are required for these.

Boat time (or need for boat trips) is a particular issue. When choosing sites for transplantation or nurseries you need to bear in mind the financial implications. If nurseries are offshore then considerable boat costs are likely. If nurseries are more than about 2–3 m deep then it is likely that SCUBA will be needed for maintenance activities (see Chapter 4, section 4.5). Boats dedicated to the restoration project need to either be hired or bought and fuel will be a key operational cost if boat travel is over a significant distance (unless sails are used). Depending on how the boat is to be utilised it will need to have certain characteristics (deck space, etc). These need to be made clear so practitioners can cost appropriately for their local situation.

The most useful approach to guide other practitioners is to provide a breakdown of cost-items with example rates and costings from a real example.

 Try to break down your costs in a way that will be helpful to others.

 Try to cost all inputs to a project, even if some are "free" in your project.

 Express labour costs as person-hours (or person-day/week/month/year multiples) taken to accomplish each defined task, so that they can be converted to any currency using local wage rates.

 Separate capital/set-up/one-off costs from operational running costs so that these can be spread over a number of years (amortised/pro-rated) if appropriate.

 Express diving needs in terms of estimated number of air-tanks required to accomplish each task.

 Express boat needs in terms of number of days of boat support required to accomplish each task.

Breaking down costs

To aid in (1) identifying cost-items, (2) estimating costs at each stage of a rehabilitation project and (3) calculating costs in a way that is useful for others, we have broken down the process into six stages. Not all stages may be appropriate for every project, thus for example, for a small project where no nursery rearing was anticipated you would omit stages 2–4. For each stage we pose a series of questions which you need to answer in estimating (if at the planning stage) or evaluating costs (if trying to assess cost-effectiveness). There are additional questions relating to planning and cost-effectiveness, that you should also consider. Previous chapters and the worked examples provided below should give some guidance as to our experience of the time (person-hours) required to perform various tasks.

For a rehabilitation project, costs might be broken down as follows:

1. Collection of source material (corals of opportunity, fragments from donor colonies, mature colonies/colony-segments about to spawn, spawning slicks).

- How many person-hours are required to collect x corals of opportunity, x fragments from donor colonies, x mature colonies/colony-segments about to spawn, an amount of spawning slick necessary to generate x competent embryos, etc.?

- What is approximate cost (US$) of any equipment needed for collecting and holding the coral source material? [This cost needs to be expressed per amount of material so that costs can be scaled.]

- Is SCUBA needed or can work be done efficiently using snorkelling?

- Is boat transport needed? (What primarily determines need? Can this be minimised?)

- Which factors are likely to contribute most to costs (in terms of both time and money)? [For example, costs will depend on location (e.g. distance of donor sites from nursery or restoration sites) and local costs of purchasing or renting a boat and scuba equipment.]

2. Setting up coral culture/nursery/hatchery facilities (in situ or ex situ nurseries, tanks, etc.)

If only a very small area (e.g. 100s m² or less) is being rehabilitated and direct transplantation of fragments is proposed, or corals are being translocated from a site threatened by development (e.g. construction or dredging) to a safer site, then material may just be held temporarily in the field, but there may be some equipment/consumable/person-hour costs associated with this.

- What are costs of equipment/consumables/staff time (person-hours) to set up nurseries/tanks? [These costs need to be expressed per amount of material which facilities can handle (e.g. per 1000 or 10,000 fragments/nubbins or per 10,000 or 100,000 newly settled coral spat) so that costs can be scaled to size of operation planned.]

- How long are these facilities likely to last and what annual inputs (on average) are likely to be required to keep facilities functional and in a good state of repair? [If a facility will last for 5 years, then costs can be spread over 5 years.]

3. Establishing collected coral material in culture/nurseries.

- What are time and consumable costs involved in setting up *x* amount of coral material (e.g. 1000 fragments or 100,000 settled coral spat) in culture? [For asexual fragments, this might include plastic pins/wall-plugs/hose-pipe/other rearing substrates, glue, cutters, etc. and person-hours to set up *x* amount of coral in an *in situ* nursery.]

4. Maintenance of corals in culture.

- What maintenance activities are required to ensure good survival of corals?

- How many person-hours are required per month/year/culture cycle to maintain material?

- What are consumable/equipment/boat/SCUBA costs involved?

- What is likely outcome if no maintenance is carried out? Is some basic level of maintenance mandatory to avoid high mortality; are some activities discretionary (i.e. their cost-effectiveness is marginal)?

5. Transfer of corals from culture/nursery/farm or source reef and attachment at the rehabilitation site.

- How many person-hours are required per *x* amount of material to transfer cultured/farmed/collected corals from nursery site or source reef to the rehabilitation site?

- What are consumable/equipment/boat/SCUBA costs per *x* amount of material?

- What factors primarily determine these costs? (e.g. distance to restoration site)

6. Maintenance and monitoring of transplants at the rehabilitation site.

- What maintenance activities and at what frequency are recommended to enhance survival of transplants? [What types of maintenance are likely to be most cost-effective? How does need for maintenance vary with environmental conditions (e.g. water quality, herbivory)?]

- How many person-hours are needed for these activities per unit area restored?

- What are likely associated consumable/boat/SCUBA costs?

Monitoring is needed both to evaluate the success/failure of your project (Chapter 2) and to allow adaptive management if things do not go according to plan (Chapter 3). More elaborate forms of monitoring are largely scientific exercises which should be separated from maintenance in costing. Maintenance contributes directly to the success of restoration and necessarily involves an element of basic "monitoring" to allow adaptive management (e.g. survival of transplants, % coral cover, presence of disease, presence of predators such as Crown-of-thorns starfish). However, more detailed scientific monitoring such as measurement of individual coral colony growth, although highly recommended, can be regarded as a separate overhead. It does not contribute directly to restoration success although it may ultimately contribute to a better understanding of reef recovery processes and thus better adaptive management of restoration projects.

- What monitoring is needed (frequency and type of monitoring) to allow (a) adaptive management (in the event that problems arise) and (b) evaluation of your reef restoration project?

- For how long does monitoring need to be carried out post-transplantation? [This will depend on the aims and objectives and the criteria adopted for evaluating the success of the project (Chapter 2).]

- What are likely costs in person-hours, boat, SCUBA, consumables per year to achieve this for *x* area of rehabilitated reef?

We now attempt to apply our 6-step procedure to a few experimental trials of reef restoration techniques with a focus on comparing the cost-effectiveness of different methods and identifying which stages are most costly.

7.3 Cost-analysis of an example reef rehabilitation project

In this section we examine the costs of a real experimental transplantation, analyse them in a spreadsheet to identify where the principal costs lie (e.g. equipment, consumables or labour; collection of material, nursery rearing or transplantation) and then explore some "What if? Scenarios" to see where efficiency gains can most effectively be made and thus where research into improved techniques should be focused.

The example that we have chosen to illustrate the process involves (1) the collection of coral source material from donor colonies, (2) construction of an *in situ* modular tray nursery (see section 4.3), (3) establishing coral fragments in the nursery, (4) maintaining the corals in the nursery over one year, (5) transferring the reared corals to rehabilitation sites and attaching them, primarily with epoxy putty, to areas of degraded reef, and (6) monitoring and maintaining the transplants over one year. The example is set in a developing country and is for a nursery that can produce around 10,000 small colonies per year of a size suitable for transplantation (say 7–10 cm for branching and 4–5 cm diameter for massive, sub-massive and encrusting corals), which is situated about 8 km by boat from a home base across sheltered water and within 4 km of the rehabilitation sites. Costs for construction would apply to fixed modular tray or lagoonal floating nurseries (Chapter 4). Nursery and rehabilitation sites were in a shallow lagoon (< 5 m deep).

For each of the six stages, costs are broken down into (a) equipment and consumables, which are costed in dollars (converted from local currency) and (b) labour, diving and boat needs, which are estimated in terms of person-hours, air tanks and boat-days respectively (Table 7.2). We present this one example in full as it provides a template which we hope will be useful to others. It both shows you the data that underpin the spreadsheet we discuss below and gives you an idea of the kind of information that is needed to cost a project fully.

Table 7.2 Breakdown of costs for a rehabilitation project using *in-situ* modular tray nurseries with capacity to produce 10,000 coral transplants per year.

1. Collection of source material – 10,000 fragments from donor colonies.

1a. Equipment/consumables needed to collect source material

Item	Unit cost	Quantity	Total cost
Chisel	$3.00	2	$6
Hammer	$4.00	2	$8
Baskets	$1.50	6	$9
Cutters	$3.50	2	$7
Total			**US$30**

1b. Labour/diving/boat time needed to collect source material

Item	Breakdown	Total
Person-hours (#)	2 people x 10 h	**20**
Air-tanks (#)	2 people x 10 tanks	**20**
Boat time (days)	6 x half-day trips	**3**

• Collection of source material may become more costly per fragment as numbers are scaled up because of the need to go further afield to find either corals of opportunity or donor colonies.

2. Setting up *in situ* modular tray nursery facilities.

2a. Equipment/consumables needed to construct c. 10,000 fragment tray nursery

Item	Total cost
PVC pipes, connectors and glue	$725
Plastic mesh	$70
Cable-ties	$140
Ropes	$75
Buoys	$380
Metal stakes (angle iron)	$75
Cement	$15
Miscellaneous	$170
Total	**US$1650**

2b. Labour/diving/boat time needed to construct c. 10,000 fragment tray nursery

Item	Breakdown	Total
Person-hours (#)	Land: 2 people x 10 days x 10 h	200
	Modular trays: 2 people x 5 days x 7 h	70
	Installing ropes and buoys: 2 people x 3 days x 10 h	60
	Nursery deployment: 4 people x 3 days x 4 h	48
	Total	**378**
Air-tanks (#)	54 tanks for deployment and rope/buoy installation	**54**
Boat time (days)	3 full-days nursery deployment and 3 full-days installing ropes and buoys	**6**

3. Establishing collected material in culture/nurseries.

3a. Equipment/consumables needed to establish c. 10,000 fragments in a tray nursery

Item	Unit cost	Quantity	Total
Cutters	$3.50	2	$7
Plastic containers (50 l)	$10.00	4	$40
Cyanoacrylate glue	$1.50	50	$75
Substrate for fragments [†]			$20
Total			**US$142**

[†]68 m plastic pipe for branching species; 9 x 22 m (198 m²) plastic mesh for submassives.

3b. Labour/diving/boat time needed to establish c. 10,000 fragments in a tray nursery

Item	Breakdown	Total
Person-hours (#)	4 people x 7 h x 63 days (Transplanting corals on trays and deploying in nursery)	**1764**
Air-tanks (#)	2 tanks per person (4) per day (63)	**504**
Boat time (days)	63 full-days of boat	**63**

• Average time to glue and transplant coral to tray = 1.5 min (250 hours for 10,000)

4. Maintenance of material in culture.

4a. Equipment/consumables needed to maintain c. 10,000 fragment tray nursery for one year

Item	Unit cost	Quantity	Total cost
Brushes	$1	2	$2
Gloves	$1	2	$2
Spare buoys			$67
Rope			$5
Cable ties			$33
Total			**US$109**

4b. Labour/diving/boat time needed to maintain c. 10,000 fragment tray nursery for one year

Item	Breakdown	Total
Person-hours (#)	2 people x 6 h (5 times/month)	**720**
Air-tanks (#)	4 tanks per visit (5 times/month)	**240**
Boat time (days)	1 full-day trip per visit (5 times/month)	**60**

• Estimates range from 2 people x 6 h x 4 times/month (576 person-hours) to 2400 person-hours. An intermediate estimate is that modular table nurseries are cleaned twice a month for 2–3 days (4–6 days/month) by two or more divers with 6–7 h per day spent cleaning the nurseries (no scientific monitoring included). The latter suggests at least 720 person-hours per year.

• The amount of maintenance needed may vary by a factor of 3 or even more from site to site depending on local water quality, abundance of herbivores to keep algae in check and abundance of predators of pests (e.g. fish that eat young *Drupella*). There was heavy fishing pressure at the example site, thus herbivorous and predatory fish were rare. The nearest aquaculture ponds and sources of nutrient rich run-off were about 3 km away from the nursery site. Estimated maintenance effort quoted here might be doubled or halved depending on the water quality at your proposed site.

5. Transfer and attachment of material from *in-situ* modular tray nursery to the restoration site.

5a. Equipment/consumables needed for transport of c. 10,000 fragments and attachment at restoration site (based on data for 1,000)

Item	Unit cost	Quantity	Total cost
Baskets	$1.00	5	$5
Plastic containers	$3.50	10	$35
		Total (transport)	**$40**
Nails			$250
Hammers			$15
Epoxy putty			$500
Wire brushes			$10
		Total (attachment)	**$775**
Total			**US$815**

5b. Labour/diving/boat time needed for cleaning/transport of c. 10,000 fragments and attachment at restoration site (based on data for 1,000)

Item	Breakdown	Total
Person-hours (#)	Cleaning/transport – 4 people x 2 h per day x 100 days	800
	Attachment – 4 people x 6 h per day x 100 days	2400
	Total	**3200**
Air-tanks (#)	Cleaning/transport – 4 tanks per day x 100 days	400
	Attachment – 8 tanks per day x 100 days	800
	Total	**1200**
Boat time (days)	Transport and attachment – 1 full-day trip for 100 days	100
	Total	**100**

6. Maintenance and monitoring of transplants at restoration site.

6a. Equipment/consumables needed to maintain/monitor 10,000 transplants

Item	Unit cost	Quantity	Total cost
No additional equipment needed			0
Total			**US$0**

6b. Labour/diving/boat time needed to maintain/monitor 10,000 transplants

Item	Breakdown	Total
Person-hours (#)	Maintenance visits 12 times/year: 2 people, full-day visit (8 h)	**192**
Air-tanks (#)	Maintenance check 12 times/year (8 tanks per survey – 4 per person)	**96**
Boat time (days)	Maintenance check visits: 12 full-day trips	**12**

• Figures above, for maintenance at transplant site, are estimates. The resources allocated allow some time for adaptive management in the event of problems being identified (e.g. COT or *Drupella* infestations) and assume little maintenance of transplants (e.g. macroalgae removal) is needed.

7

The breakdown in Table 7.2 provides the basic data on equipment/consumables (cost in US$), time input by personnel (person-hours), number (#) of air-tanks and boat-time in days that were used at each stage (shown in green type) in Figure 7.1.

A set of local wage rates are set up in Area A of the spreadsheet (Figure 7.1) in order to convert person-hours to a US$ value that can be compared between projects and used in cost-effectiveness or benefit-cost analysis. The term "wage" is used loosely to include salary, stipend or any payment for labour. We use three different rates for three different skill levels. These are based on local rates in the Philippines but are primarily for illustration. You need to be realistic in terms of wage rates, remembering that personnel need an adequate level of skills, e.g. ability to work whilst SCUBA-diving. Monthly rates are converted to hourly ones using Table 7.1. For the tasks at each stage in the first column, an hourly rate is assigned based on the mix of skills required. Thus for stages 4 and 6, trained local manual labour at the lowest wage rate is adequate, whereas for stage 2, 10% of person-hours are assigned to the highest skill level (expert advice) with the remaining 90% split equally between local "trained educated" and "trained manual" labour.

To translate the number of air tanks required and boat needs for the project into US$, local rates for tank fills and boat hire are inserted in Area B of the spreadsheet. Most projects are likely to be long term (at least 3 years), given the time course of natural reef recovery, and it is likely to be cheaper to purchase diving equipment rather than hire it. This is examined in Area C of the spreadsheet where the cost of daily hire is compared to the cost of purchasing a full-set of diving equipment. In the breakdown of costs, the researchers indicated that the majority of work days (169 of 244) would require 4 divers. Thus the cost of purchasing 4 complete sets of diving gear (at local prices) is compared to the cost of hiring throughout the project. In the example, it is considerably cheaper to buy. For flexibility, the diving equipment cost is kept separate from the other costs and is only incorporated into the total cost at the *"What if? Scenario"* stage.

In the example, all capital equipment is given a life of three years with some allowance being made for maintenance. Thus the US$ 3,200 required to purchase the necessary diving gear (4 sets) is considered as an annual cost of US$ 1,067. You might feel that some items would last for longer, say 5 years, in which case you could spread (pro-rate) the capital cost over a 5-year period, rather than over a 3-year period as we have done here. However, in comparing costs between projects, the same period of amortisation should be used in each. Similarly the cost of hiring a boat (and driver) and the cost of fuel each day (which in the example amounts to only $30/day) can be compared to the cost of buying a project boat and running it. The decision as to which is likely to be the most sensible option needs to be discussed at the project planning stage (Chapter 2).

Capital equipment that will last for several years should be identified and separated out from consumables. In the example, over 80% of the nursery construction costs (US$ 1,380) are considered to be structural items that will last for three years and thus these costs are spread over 3 years (Area D in Figure 7.1). This allows the cost of equipment and consumables to be expressed as an annualised cost which can be used when calculating the cost per transplant. Thus for each crop of 10,000 transplants produced in the coral nursery, only US$ 460 (US$ 1,380 /3) is attributed to costs of nursery construction materials (i.e. <5 cents/transplant).

Cost analysis

The final column in Figure 7.1 examines the percentage of the annualised costs relating to each stage. This shows that using the methods chosen, the most expensive stage in the project cycle is the transfer of reared coral colonies from the nursery and their attachment at the rehabilitation sites. This task accounted for an estimated 50% of costs. The second most time-consuming activity was stocking the modular tray nursery with the transplants which accounted for around 26% of costs. Thus, it is at these two stages where increased efficiency can offer most gains.

Another important outcome of the analysis is that the capital cost of materials to build the nursery represent only about 5% of the overall annual costs once the project is fully up and running. Thus, if you try to make savings by using cheap materials that may not last, this is likely to be a false economy (a point highlighted in Chapter 4). It clearly makes sense to make sure the nursery structure is as robust as possible to minimise the risk that it will be damaged or destroyed by a storm.

In the example over half the total costs relate to the labour (person-hours) required to operate the nursery and carry out transplantation. The other major item is the cost of SCUBA-diving and boat time. In setting up a project you could reduce the amount of boat time needed by minimising distances that need to be travelled (e.g. distances from home base to coral nursery or from coral nursery to proposed rehabilitation site), but often convenient sites may not be suitable for coral nurseries and reefs in need of rehabilitation may not necessarily be close to your home base. Thus, most projects are likely to be constrained by local geography and ecological and social realities. However, if the amount of time spent doing the various tasks can be reduced, then the amount of SCUBA time and boat time will also be reduced. The example suggests that there are two disproportionately time-consuming stages that require research and development of new techniques. Firstly, how to stock coral nurseries more cost-effectively; secondly, how to transplant corals to degraded reefs more cost-effectively. Various researchers around the world are currently working on ways of doing both and in the next section we examine how advances in methodology might affect cost-effectiveness in a series of *"What if? Scenarios"*.

Asexual rearing to produce 10,000 coral fragments/year using *in-situ* modular tray nurseries

Local wage rates	Monthly	Monthly/160 Hourly
Skill level 1 (highest) salary - e.g. scientific adviser/expert	$900	$5.63
Skill level 2 (medium) salary - e.g. trained educated local	$560	$3.50
Skill level 3 (lowest) salary - e.g. trained manual labour	$210	$1.31

A

Local rates (air fills and boat hire)
air per tank $2   boat per day $30

B

Task	Equipment/ consumables USD	Time input by personnel Person-hours	Rate	USD	# air-tanks	USD	Boat time days	USD	Total cost/yr USD	% total cost/yr
1. Collection of source material for 10,000 fragments	$30	20	$3.48	$70	20	$40	3	$90	$230	1%
2. Setting up 10,000 fragment tray nursery	$1,650	378	$2.73	$1,031	54	$108	6	$180	$2,049	7%
3. Establishing 10,000 fragments in nursery-culture	$142	1764	$2.41	$4,245	504	$1,008	63	$1,890	$7,285	26%
4. Maintenance of material in *in-situ* culture (1 year)	$109	720	$1.31	$945	240	$480	60	$1,800	$3,334	12%
5. Transfer to and attachment of 10,000 juveniles at restoration site	$815	3200	$2.41	$7,700	1200	$2,400	100	$3,000	$13,915	50%
6. Monitoring and maintenance at restoration site (1 year)	$0	192	$1.31	$252	96	$192	12	$360	$804	3%
Annualised total (nursery construction cost split over 3 years)	**$1,826**	**6274**		**$14,242**	**2114**	**$4,228**	**244**	**$7,320**	**$27,616**	**100%**
Capital equipment - costs split over 3 years	$1,380									
Diving equipment - costs split over 3 years	$3,200									

D

Diving gear	Per set	Rate	Persons	Total	
Daily hire		$20	2 or 4	$16,520	
Purchase		$800	4	$3,200	Cheaper to buy 4 sets

Boat days with 2 divers		75
Boat days with 4 divers		169

C

What if? Scenarios

1.						
%survival in nursery	100%	95%	90%	85%	80%	75%
a) Cost per 1-year fragment in nursery	$1.40	$1.47	$1.55	$1.64	$1.75	$1.86
b) Cost per 1-year transplant at restoration site	$2.79	$2.93	$3.10	$3.28	$3.48	$3.72
2.						
% survival at restoration site (over one year)	100%	95%	90%	85%	80%	75%
Cost per 2-year transplanted colony at restoration site	$2.87	$3.02	$3.19	$3.37	$3.59	$3.82

Figure 7.1. Analysis of costs for a rehabilitation project using *in-situ* modular tray nurseries with capacity to produce 10,000 coral transplants per year. The green horizontal line between stages 4 and 5, divides the costs of rearing the coral fragments in a nursery from those of transplanting them to a degraded reef.

What if? Scenarios

Given the cost assumptions laid out in Areas A, B, C and D of the spreadsheet example shown in Figure 7.1 (available on-line at www.gefcoral.org/Targetedresearch/Restoration/Informationresources/Costingrestoration/ as Example_7.1a.xls), the cost of rearing a coral fragment for one year in the *in situ* modular tray nursery can be calculated based on various survival rates. In a well-maintained nursery, you might expect 10–15% mortality over a year plus maybe 5% loss due to detachment (section 4.4). Thus survival should be around 80–85%. For our first *What if? Scenario*, we have looked at the effect of changing survival rate from 75% (pessimistic scenario) to 100% (optimistic and unlikely scenario) on costs of (a) rearing each healthy transplant (Cost per 1-year fragment in nursery) and (b) transporting and securely attaching that transplant on a degraded reef (Cost per 1-year transplant at restoration site). The analysis suggests that each transplant-ready, healthy small colony reared in the nursery will have cost from US$ 1.40–1.86 to rear. However, owing to the costs entailed in transplantation, once on the degraded reef costs rise to US$ 2.79–3.72 per transplant (Figure 7.1). (Although costs are expressed to the nearest cent, when interpreting the outputs you should bear in mind that this is a false precision and treat estimates as accurate to the nearest dollar given the stated assumptions.) For our second *What if? Scenario*, we examine the costs after one year at the rehabilitation site based on survival rates ranging from 75% to 100% (several studies indicate that ~80% survival over one year is achievable). Now the costs per surviving transplant rise to US$ 2.87–3.82 (Figure 7.1). Since the assumptions and person-hours spent on each stage are transparent, you can readily change these in the spreadsheet.

Research has led to improved techniques since the nurseries in the example were established. We will now examine how the benefits of this research might affect estimated costs and cost-effectiveness. Analysis of a range of projects (Box 2.4), indicates that the rate of attachment of colonies to the reef using epoxy averages about 4–5 colonies per person-hour (as in the example). This appears slow but includes all the peripheral activities involved, not just the actual time spent underwater fixing colonies to the degraded reef. Efficiency gains can be made in a number of ways. For branching species, the time needed both to establish fragments in a nursery and deploy to the reef after a period of rearing can be reduced by using substrates such as wall-plugs (section 4.4). Also, being able to attach fragments to rearing-substrates without using glue, can reduce the time needed to stock the nursery. Similarly, being able to slot reared colonies into pre-drilled holes on the reef rather than attach them using epoxy can triple or quadruple the numbers of transplants than can be deployed per unit time. Further, careful use of "environmentally friendly" anti-fouling paint can reduce the time needed for maintenance (section 4.5).

Given these potential efficiency gains, what effect on costs would a 30% decrease in the time needed to stock the nursery, a 50% decrease in maintenance costs and a ~55% decrease in time needed to clean, transport and attach the transplants to the reef have? For the latter activity, we do not envisage that the time spent cleaning the nursery reared transplants (removing algae, sessile invertebrates, corallivorous snails, etc.) and transporting them can be reduced but we assume a *fourfold* improvement in the rate at which they are attached to the degraded reef. The results are summarised in Figure 7.2 below.

Task	Total cost/yr USD	% total cost/yr
1. Collection of source material for 10,000 fragments	$230	1%
2. Setting up 10,000 fragment tray nursery	$2,049	12%
3. Establishing 10,000 fragments in nursery-culture	$5,131	31%
4. Maintenance of material in *in-situ* culture (1 year)	$1,722	10%
5. Transfer to and attachment of 10,000 juveniles at restoration site	$6,584	40%
6. Monitoring and maintenance at restoration site (1 year)	$804	5%
Annualised total (nursery construction cost split over 3 years)	**$16,519**	100%

What if? Scenarios

		100%	95%	90%	85%	80%	75%
1.	% survival in nursery	100%	95%	90%	85%	80%	75%
a)	Cost per 1-year fragment in nursery	$1.02	$1.07	$1.13	$1.20	$1.27	$1.36
b)	Cost per 1-year transplant at restoration site	$1.68	$1.77	$1.86	$1.97	$2.10	$2.24
2.	% survival at restoration site (over one year)	100%	95%	90%	85%	80%	75%
	Cost per 2-year transplanted colony at restoration site	$1.76	$1.85	$1.95	$2.07	$2.20	$2.34

Figure 7.2. Summary of the effects on costs of assuming major efficiency gains at stocking, maintenance and attachment stages of the example project. Full spreadsheet is available at: www.gefcoral.org/Targetedresearch/Restoration/Informationresources/Costingrestoration/ as Example_7.1b.xls.

As can be seen by comparing Figure 7.2 with 7.1, the total annualised cost of the example project has decreased by almost 40% as a result of the efficiency gains. The relative costs of each of the stages have remained as before except that the cost of setting up the nursery has assumed greater importance, now representing about 12% as opposed to 7% of total costs. In terms of costs per nursery-reared colony ready to be transplanted, the analysis suggests that a unit cost of around US$ 1 is achievable with existing techniques (given the assumptions in Areas A–D of Figure 7.1). This rises to around US$ 2 for a transplant securely attached to a degraded reef. Approximately 50% of total costs remain as labour (staff time).

In the next section we compare costs of rearing fragments asexually with rearing sexually produced coral spat.

> ### Box 7.1 Choosing an endpoint for comparison of methods
>
> You can calculate unit costs for endpoints at various stages in the project cycle in order to compare different methods. The important thing is that comparisons are valid, that is, the assumptions in Areas A–D of the spreadsheet (Figure 7.1) are the same and the same endpoint is used in any comparison between methods. Comparing techniques at early stages (e.g. survival after 3 months in a nursery) is not particularly useful as short term success can often be short-lived. If you wish to compare different nursery rearing methods then costs per live coral at a size large enough to be transplanted would be a sensible endpoint. However, costs per transplant successfully deployed to the reef would be a better endpoint because our analysis shows that the next stage is critical in terms of overall costs (accounting for 40–50% of costs depending on assumptions – Figures 7.1 and 7.2). What is the point of rearing corals very cheaply if you cannot deploy them effectively?
>
> The next test is the survival of the transplants. You could assess this annually but it is unclear when to stop (5 years? 10 years?) and also unclear what constitutes "good" survival and what "poor". Natural survival rates vary from species to species, site to site and through time at a given site due to stochastic disturbances. Comparing survival of transplants with natural populations of the same species (Ch. 8: Case study 5) is one approach but involves a substantial monitoring overhead. As a way of forcing "closure" we recommend *the cost per reproductively mature colony* at the transplant site as a scientifically defensible benchmark for comparison between methods. (If transplants never reach this stage, it is unclear what has usefully been achieved, although there are arguments that creating temporary topographic diversity in the form of dead coral has some ecological engineering benefits. With good cost information, these could be examined by means of benefit-cost analysis[2].) Once the transplanted corals are spawning or releasing brooded larvae at the rehabilitation site, you are in a strong position to argue that some measure of effective ecological restoration has been achieved. This seems a defensible endpoint, which is likely to be reached over a time-scale (several years) comparable to that of natural recovery and runs a relatively low risk of premature comparison (i.e. it is not unusual for rehabilitation projects to be apparently successful after one year, but to have failed by two years). However, the final link in the chain of recovery (for the coral community) is whether coral larvae are able to settle and survive at the site. If careful inspection of reef substrate shows that coral recruitment is occurring (see Box 2.2 for discussion of recruitment limitation), then you can be sure that the full coral life-cycle (Figure 5.1) is being completed at the rehabilitation site. A self-sustaining coral community appears the most ecologically defensible goal.

7.4 Comparing cost scenarios

Rearing of sexually produced coral larvae is at a more experimental stage than rearing coral fragments and requires greater expertise and more specialist equipment (Chapter 5). We have broken down the costs of two slightly different methods of rearing coral spat and will compare them to each other and to the example of asexual rearing (Table 7.2, Figure 7.1). The original pilot studies were carried out in Palau and the Philippines respectively but we have inserted the same wage rates, air and boat and diving gear costs, and amortisation of capital equipment assumptions as we did for the first example so that valid comparisons can be made. If you wish to examine them, the breakdowns of costs for each of these two projects are available at: www.gefcoral.org/Targetedresearch/Restoration /Informationresources/Costingrestoration/ as Example_7.2.pdf and Example_7.3.pdf. The first project involved spawning corals in tanks, settling the larvae on tiles and then rearing these for one year inside cages in an *in-situ* nursery in co-culture with *Trochus*. The second project differed from the first in that the coral spat were settled onto special substrates ("coral plug-ins" – see section 5.9) and reared in a semi-caged *in situ* nursery prior to deployment to the reef. Both were early pilot experiments from which many lessons were learnt. The first set out to explore the feasibility and costs of producing 2000–2500 juvenile corals per year for restoration; the second focused on producing about 1000 coral plug-ins per year for restoration (each with at least one live coral surviving). Thus, both were small-scale.

For corals reared from larvae, evaluating unit costs and cost-effectiveness is harder than for the asexual nurseries as the starting point is less well defined. For the asexual nurseries we knew the number of fragments with which we had stocked the nursery. For larval rearing we are likely to start with several hundred thousand to millions of embryos but will not know how many until the corals spawn. However, at settlement time we can make an estimate of the number of coral polyps settled on our substrates (section 5.8 and 5.9). This will vary with each batch of corals but provides a starting point from which survival can be monitored and unit costs calculated. If larvae are carefully looked after, then 70–90% should survive the few days required until they are competent to settle and metamorphose into tiny coral polyps. With good husbandry you should be able to settle 60–80% of the larvae you start with onto "conditioned" substrates (see Box 5.5) on which they can then be reared. The critical stages are the survival of the larvae in *ex situ* and *in situ* nurseries until ready for transplanting to the degraded reef, and then the survival once outplanted to a reef. In nature maybe one in a million larvae survives to become an adult coral colony. If the survival rate can be increased by four orders of magnitude to one in 100, then there is the potential to generate 10,000 juveniles from a million larvae. Key questions are what sort of survival can be achieved using larval culture techniques and at what point might these become cost-effective?

Rearing coral spat in co-culture with Trochus *in in-situ cage nurseries*

Figure 7.3 presents an analysis of the costs of the first pilot experiment (coral spat reared *in-situ* nursery in co-culture with *Trochus*), which set out to rear 2000–2500 juveniles per year. Almost 70% of annualised costs relate to the building of the *in situ* cage nursery, rearing of the larvae and stocking of the nursery with recently settled corals. Equipment and consumable costs are much more important in terms of total inputs than for the asexual nursery (amounting to about two-thirds of annualised costs), due to the technical nature of the rearing work. Approximately 168,000 larvae were settled on tiles and we take this as the starting point for the *What if? Scenarios*. Actual survival over the first year was approximately 0.5% (due primarily to poorer water quality than anticipated at original nursery site and sub-optimal maintenance). This seems low but natural survival in the wild would have likely been around 0.001%. Thus in the pilot experiment, each live juvenile coral has cost about US$ 8.50 to produce or over US$ 9 once transplanted to a reef. In comparison to the asexual rearing (Figure 7.1) this is about seven times as costly, moreover, at this stage the sexually reared coral colonies are likely to be smaller and thus more vulnerable than coral fragments reared in a nursery for one year. (For the narrow range of first year survival rates, costs have been assumed to change only marginally for this analysis.)

For the actual experiment, wage rates, air and dive boat costs were significantly greater than those used for the purposes of comparison. Indeed the total annualised costs were in reality slightly more than twice those on the spreadsheet (US$ 15,738 as opposed to US$ 7,415 in the example). However, because the researchers provided details of effort in terms of person-hours, numbers of air-tanks and boat-days, we have been able to restate the costs using the same assumptions as for the first example and thus validly compare them with work carried out in the Philippines.

How can efficiency gains be made? If you examine the effects of improving the first year survival from 0.5–1.5%, you can see that with 1.5% survival at 1-year post-settlement, the experiment would have achieved its goal of 2500 juveniles per year for the settlement achieved (168,000 coral spat). The spreadsheet shows that small incremental gains in first year survival markedly reduce costs and thus improving one year survival to 1–2% is clearly one area on which to focus in order to bring unit-costs down to those achieved by the asexual rearing techniques. However, as for asexually produced transplants, the survival of juveniles once transplanted to the reef is also critical. In this case, less than 20% of corals survived the first year after transplantation to the reef, possibly due to parrotfish (*Bolbometopon muricatum*), triggerfish and boxfish predation. Even with 20% survival, the cost-analysis shows that each surviving two-year old colony would have cost almost US$ 50 to generate (given the 0.5% survival in the nursery) – a few weeks wages for a Marine Protected Area guard in some parts of the world.

Settlement of over 500,000 coral spat is now regularly achieved in larval rearing trials in tanks. What if this level of settlement had been achieved for the pilot experiment? We assume that stage 1 and 3 costs would be little changed but increase costs of other stages on a *pro rata* basis to arrive at unit costs in Figure 7.4. The *What if? Scenarios* show that to achieve something close to a US$ 5 two-year colony, nursery survival would need to improve to at least 1.5–2% and 1-year survival of transplants on the reef would need to be at least 50% (using these techniques). The analysis provides some useful medium-term goals for researchers. Perhaps the most tantalizing outcome is the how the achievement of tiny decreases in early mortality have such potential to generate large numbers of juvenile corals. If this potential can be harnessed, costs could decrease significantly.

Larval rearing (*In situ* cage-culture to produce 2000-2500 juvenile corals per year in 10 cages)

Local wage rates	Monthly	Monthly/160 Hourly
Skill level 1 (highest) salary - e.g. scientific adviser/expert	$900	$5.63
Skill level 2 (medium) salary - e.g. trained educated local	$560	$3.50
Skill level 3 (lowest) salary - e.g. trained manual labour	$210	$1.31

Local rates (air fills and boat hire)
air per tank $2   boat per day $30

Item	Equipment/ consumables USD	Time input by personnel Person-hours	Rate	USD	# air-tanks	USD	Boat time days	USD	Total cost/yr USD	% total cost/yr
1.1. Surveys to predict dates of spawning	$14	20	$4.56	$91	20	$40	5	$150	$295	4%
1.2. Collection of portions of 10-15 gravid colonies of 3 species	$125	4	$4.56	$18	4	$8	0.5	$15	$166	2%
2.1. Construction of 10-cage nursery with 320 tiles (lifetime c. 3	$4,440	54	$2.41	$130	18	$36	3	$90	$1,736	23%
2.2. Annual input of 320 tiles, tile holders, cage repair, etc.	$1,120	0		$0	0	$0	0	$0	$1,120	15%
3. Establishing material in cage-culture	$3,300	134	$4.56	$611	0	$0	0	$0	$2,261	30%
4. Maintenance of material in cage-culture (1 year)	$140	162	$1.31	$213	162	$324	13.5	$405	$1,082	15%
5. Transfer of juveniles to restoration site	$420	34	$3.48	$118	24	$48	1.5	$45	$631	9%
6. Monitoring and maintenance (4 times in year)	$10	16	$1.31	$21	16	$32	2	$60	$123	2%
Annualised total (nursery construction cost split over 3 years)	**$4,959**	**424**		**$1,203**	**244**	**$488**	**25.5**	**$765**	**$7,415**	**100%**
Capital equipment - costs split over 3 years	$6,915									
Diving equipment - costs split over 3 years	$1,600									

Diving gear	Per set	Rate	Persons	Total	
2-yr cycle	Daily hire	$20	2-10	$1,240	Cheaper to hire over 2-yr cycle
	Purchase	$800	2	$1,600	Cheaper to buy 2 sets over 3 years

Number of coral polyps settled   168,000

Boat days with 2 divers	21.5
Boat days with 3 divers	3
Boat days - 5 divers	1

What if? Scenarios

% survival at one year (of 168,000 settled polyps)	0.50%	0.75%	1.00%	1.25%	1.50%
Survivors after 1-year cage-culture	840	1260	1680	2100	2520
Cost per 1-year juvenile in cage culture	$8.56	$5.71	$4.28	$3.43	$2.85
Cost per 1-year juvenile outplanted	$9.32	$6.21	$4.66	$3.73	$3.11

Cost per 2-year colony

Survival rate (%) from 1 to 2-years old					
70	$13.52	$9.01	$6.76	$5.41	$4.51
60	$15.77	$10.51	$7.88	$6.31	$5.26
50	$18.92	$12.62	$9.46	$7.57	$6.31
40	$23.65	$15.77	$11.83	$9.46	$7.88
30	$31.54	$21.03	$15.77	$12.62	$10.51
20	$47.31	$31.54	$23.65	$18.92	$15.77
15	$63.08	$42.05	$31.54	$25.23	$21.03
10	$94.62	$63.08	$47.31	$37.85	$31.54
5	$189.24	$126.16	$94.62	$75.70	$63.08

Figure 7.3. Analysis of costs for a rehabilitation project using *in-situ* cage culture to produce 2000–2500 juvenile corals per year. The green horizontal line between stages 4 and 5, divides the costs of rearing the coral spat in a nursery from those of transplanting them to a degraded reef. Spreadsheet available at: www.gefcoral.org/Targetedresearch/Restoration/Informationresources/Costingrestoration/ as Example_7.2a.xls.

% survival at one year (of 500,000 settled polyps)	0.5%	1.0%	1.5%	2.0%	2.5%
Survivors after 1-year cage-culture	2500	5000	7500	10000	12500
Cost per 1-year juvenile in cage culture	$2.88	$1.91	$1.69	$1.59	$1.52
Cost per 1-year juvenile outplanted	$3.63	$2.66	$2.45	$2.34	$2.27
Cost per 2-year colony					
Survival rate (%) from 1 to 2-years old					
80	$4.68	$3.47	$3.20	$3.07	$2.99
70	$5.35	$3.97	$3.66	$3.51	$3.41
60	$6.24	$4.63	$4.27	$4.09	$3.98
50	$7.49	$5.56	$5.13	$4.91	$4.78
40	$9.37	$6.95	$6.41	$6.14	$5.97
30	$12.49	$9.27	$8.54	$8.18	$7.96
20	$18.73	$13.90	$12.81	$12.27	$11.95
10	$37.46	$27.80	$25.63	$24.54	$23.89

Figure 7.4. Summary of the effects of assuming settlement of 500,000 (rather than 168,000 larvae) on unit costs of juvenile coral production.

Pilot experiments show that, with care (see sections 5.5–5.9), hundreds of thousands of coral polyps can be settled onto substrates in tanks. However, even if survival rates can be improved several fold, *What if? Scenarios* suggest that unit costs still remain relatively high because of the costs of deploying the juveniles at the rehabilitation site. Thus, as for the asexual rearing of fragments (section 4.4), there is a need for cheap substrates, onto which larvae can be settled, that can be easily attached to degraded reefs once the corals are large enough.

Settling coral spat on special substrates ("coral plug-ins") and rearing in a semi-caged *in situ* nursery

Figure 7.5 presents an analysis of the costs of the second type of pilot experiment where coral larvae were settled on coral plug-ins (see section 5.9) and reared in a semi-caged *in situ* nursery prior to deployment to the reef. The aim was to produce about 1000 coral plug-ins per year for restoration (each with at least one live coral surviving). Here the starting point is the number of the specially designed and conditioned substrates (10 mm wall-plugs with 20 mm diameter x 15 mm deep cement heads = coral plug-ins)

that are used. Each one of these that has a live coral surviving is considered a success. In contrast to the previous method, only about 30% of total annualised costs were for equipment and consumables. As for the asexual nursery rearing about 50% of costs were for labour (Figure 7.5 - facing page).

Although it was possible to get coral larvae to settle on almost all the plug-ins, after one year the proportion with live corals surviving was generally less than 40% (leaving <400 plug-ins with corals), thus outplants were costing at least US$ 15 each (Figure 7.5), an order of magnitude more expensive than the asexual rearing method of Figure 7.1. Moreover 1-year survival after transplantation did not exceed 50%, such that costs per 2-year colony were at least US$ 30. However, several hundred larvae were settling on each plug-in such that densities were probably detrimentally high and the larvae could have been settled on far more substrates at almost no extra effort. We now examine how scaling up to around 5000 coral plug-ins might improve cost-effectiveness. The scaling of the costs was complex and the outcomes of the *What if? Scenario* are shown in Figure 7.6. (below).

# plug-ins with 1+ live colonies after one year	3000	4000	5000
Cost per 1-year juvenile in nursery	$2.38	$1.78	$1.43
Cost per 1-year juvenile outplant	$4.65	$3.49	$2.79
Cost per 2-year colony			
Survival rate (%) from 1 to 2-years old			
80	$6.49	$4.87	$3.89
70	$7.41	$5.56	$4.45
60	$8.65	$6.49	$5.19
50	$10.38	$7.79	$6.23
40	$12.98	$9.73	$7.79
30	$17.30	$12.98	$10.38
20	$25.95	$19.46	$15.57
10	$51.90	$38.93	$31.14

Figure 7.6. Summary of the effects of settling larvae on 5000 coral plug-ins rather than 1000 on unit costs of juvenile coral production. Full spreadsheet available at: www.gefcoral.org/Targetedresearch /Restoration/Informationresources/ Costingrestoration/ as Example_7.3b.xls.

Larval rearing (*In situ* culture to produce 1000 coral plug-ins with juvenile corals)

Local wage rates	Monthly	Monthly/160 Hourly
Skill level 1 (highest) salary - e.g. scientific adviser/expert	$900	$5.63
Skill level 2 (medium) salary - e.g. trained educated local	$560	$3.50
Skill level 3 (lowest) salary - e.g. trained manual labour	$210	$1.31

Local rates (air fills and boat hire)

air per tank $2   boat per day $30

Item	Equipment/ consumables USD	Time input by personnel Person-hours	Rate	USD	# air-tanks	USD	Boat time days	USD	Total cost/yr USD	% total cost/yr
1.1. Surveys to predict dates of spawning	$14	32	$4.56	$146	8	$16	2	$60	$236	4%
1.2. Collection of portions of up to 24 gravid colonies	$240	16	$4.56	$73	4	$8	1	$30	$351	6%
2. Construction of *in situ* nursery for 1000 plug-ins	$65	80	$2.41	$193	2	$4	0.5	$15	$277	5%
3. Establishing material in *in-situ* semi-caged culture	$2,962	212	$4.56	$967	8	$16	1	$30	$2,189	39%
4. Maintenance of material in *in-situ* culture (1 year)	$24	96	$1.31	$126	24	$48	6	$180	$378	7%
5. Transfer of juveniles to restoration site	$268	320	$3.48	$1,113	80	$160	10	$300	$1,841	32%
6. Monitoring and maintenance (monthly for 1 year)	$0	96	$1.31	$126	48	$96	6	$180	$402	7%
Annualised total (land-based hatchery costs split over 3 years)	**$1,786**	**852**		**$2,744**	**174**	**$348**	**26.5**	**$795**	**$5,673**	**100%**
Capital equipment - costs split over 3 years	$2,680									
Diving equipment - costs split over 3 years	$1,500									

Diving gear	Per set	Rate	Persons	Total	
2-yr cycle	Daily hire	$20	2 or 4	$1,500	Cheaper to hire at this scale
	Purchase	$800	4	$3,200	Cheaper to buy if scale up

Boat days with 2 divers	15.5
Boat days with 4 divers	11

What if? Scenarios

# plug-ins with 1+ live colonies after one year	200	400	600	800
Cost per 1-year juvenile in nursery	$19.65	$9.83	$6.55	$4.91
Cost per 1-year juvenile outplant	$28.86	$14.43	$9.62	$7.21
Cost per 2-year colony				
Survival rate (%) from 1 to 2-years old				
70	$44.10	$22.05	$14.70	$11.02
60	$51.45	$25.72	$17.15	$12.86
50	$61.73	$30.87	$20.58	$15.43
40	$77.17	$38.58	$25.72	$19.29
30	$102.89	$51.45	$34.30	$25.72
20	$154.34	$77.17	$51.45	$38.58
15	$205.78	$102.89	$68.59	$51.45
10	$308.67	$154.34	$102.89	$77.17
5	$617.34	$308.67	$205.78	$154.34

Figure 7.5. Analysis of costs for a rehabilitation project using *in-situ* semi-caged culture of coral spat settled on approximately 1000 coral plug-ins. The green horizontal line between stages 4 and 5, divides the costs of rearing the coral spat in a nursery from those of transplanting them to a degraded reef. Spreadsheet available at: www.gefcoral.org/Targetedresearch/Restoration/Information resources/Costingrestoration/ as Example_7.3a.xls.

Despite the extra culture facility, labour and maintenance costs involved, the analysis shows that unit costs are predicted to fall considerably as a result of this more effective use of the existing spawn. Recent experiments that have extended initial *ex situ* rearing times prior to transfer to *in situ* nurseries have increased the percentage of plug-ins with surviving corals at 6 months to 70–80% so a US$ 5 juvenile coral reared by this method appears attainable. The analyses show that rearing corals from sexually produced larvae remains considerably more expensive than rearing coral fragments but that there is huge potential for the former given that many hundreds of thousands to a few million larvae can be obtained from a few coral colonies. The key lies in utilising the settled larvae efficiently and being able to deploy surviving juvenile corals to the reef in a cost-effective way.

Conclusion

We hope that the above goes some way to providing a comprehensive costing framework that can be applied to future reef restoration schemes and that the three detailed breakdowns of all cost components in a consistent format and associated spreadsheet analyses will assist this.

The analyses of costs and comparisons of methods provide a rough guide to the relative costs of asexual and sexual rearing techniques and indicate why we have stressed that the sexual rearing methods remain experimental.

In terms of planning large-scale reef rehabilitation projects that use nursery rearing to reduce collateral damage to healthy reefs, we hope that the detailed breakdown of costs will allow practitioners to appreciate the sorts of inputs that are needed and encourage others to publish the costs of their reef rehabilitation projects in a way that will allow comparison. The spreadsheets can be readily adapted for use as a planning tool by inserting local wage rates and other costs.

References

1. Spurgeon, J.P.G. (2001) Improving the economic effectiveness of coral reef restoration. *Bulletin of Marine Science*, 69, 1031-1045.

2. Spurgeon, J. (1998) The socio-economic costs and benefits of coastal habitat rehabilitation and creation. *Marine Pollution Bulletin*, 37, 373-382.

Chapter 8

Learning lessons from past reef-rehabilitation projects

Overview of case studies

Spatial scales of restoration and costs per hectare

Lessons learnt from the case studies

Synopses of 10 case studies

Alasdair Edwards, Sandrine Job and Susan Wells

8.1 Introduction

The aim of this chapter is to see what lessons can be learned from the successes and failures of past and on-going reef restoration projects. These lessons can help guide project managers, scientists and other individuals interested in the restoration of damaged reefs, and provide information on suitable methods and resources required for the successful realization of such projects. They complement the information given in previous chapters of the manual.

Information for the case studies was collected through a questionnaire that was mailed to about 40 people worldwide who were known to be involved in reef rehabilitation projects. The questionnaire requested information about all the various stages of a restoration project, including aims and objectives, the general context of the project, methods used, monitoring strategy, ecological outcomes, human and financial resources required, socio-economic impact, and lessons learnt as perceived by those involved in the projects. A total of 22 questionnaires were returned. The information from these was collated in a standardised format and has been used to inform both this chapter and other chapters in the manual. Ten case studies with diverse objectives (Figure 8.1), for which respondents provided detailed information on most aspects requested, are summarised in this chapter. Expanded versions of these and eventually other case studies are being made available as downloadable PDF documents at: www.gefcoral.org/Targetedresearch/ Restoration/Informationresources/Casestudies/. We hope these will be a useful resource for those embarking on reef rehabilitation projects.

Figure 8.1. Locations of the 10 case studies summarised in this chapter. 1 Indonesia, 2 Tuvalu, 3 Maldives, 4 Thailand, 5 Israel, 6 Puerto Rico, 7 Japan, 8 Fiji, 9 Mexico, 10 Philippines.

8.2 Overview of case studies

Reef restoration projects are being undertaken in tropical seas worldwide and in response to our questionnaire we received case studies from the Atlantic Ocean (Belize, Florida, Mexico, Puerto Rico), Pacific Ocean (Fiji, Indonesia, Japan, Malaysia, New Caledonia, Philippines, Thailand, Tuvalu, Vietnam) and Indian Ocean (Israel, Kenya, La Réunion, Maldives). Five reef restoration projects (from Fiji, French Polynesia, La Réunion, Mayotte and New Caledonia) have already been summarised in our companion volume *Reef Restoration Concepts and Guidelines*[1]. The ten case studies summarised below are as follows:

1. Substrate stabilisation to promote recovery of reefs damaged by blast-fishing (Komodo National Park, Indonesia),

2. Transplantation of coral colonies to create new patch reefs on Funafuti Atoll (Tuvalu),

3. Transplantation of coral fragments and colonies at tourist resorts using coated metal frames as a substrate (Maldives),

4. Use of artificial substrates to enhance coral and fish recruitment (Phuket, Thailand),

5. Transplantation of nursery reared corals to a degraded reef at Eilat (Israel),

6. Re-attachment of broken fragments of *Acropora palmata* following a ship grounding (Mona Island, Puerto Rico),

7. Coral transplantation, using ceramic coral settlement devices, on reefs damaged by bleaching and *Acanthaster planci* (Sekisei Lagoon, Japan),

A coral farm in Uluibau tabu area off Moturiki Island, Fiji. The corals have been grown from fragments over 2 years (S. Job).

Reef framework and large Diploria strigosa *colonies crushed by the grounding of the M/V* Fortuna Reefer *at Mona Island, Puerto Rico (A. Bruckner).*

8. Transplantation of corals to a traditional no-fishing area affected by coral bleaching (Moturiki Island, Fiji),

9. Transplantation of coral fragments onto artificial reefs at a hurricane-damaged site (Cozumel, Mexico), and

10. Rehabilitation of a reef damaged by blast-fishing by stabilizing rubble using plastic mesh (Negros Island, Philippines).

All case studies derive from areas which are under either formal or informal management (e.g. national parks, marine protected areas, *tabu* areas, resort islands, or under local community supervision), such that local anthropogenic impacts are largely under control. As emphasised in earlier chapters, effective management of these impacts needs to be in place before you attempt active rehabilitation. The site with perhaps the least controls (Lofeagei Reef, Tuvalu) was susceptible to local anthropogenic impacts (rubbish, run-off) and suffered sudden unexplained mortality of corals between 12 months and 15 months after transplantation. The objectives of the projects which generated the case studies can be grouped into various categories:

- As mitigation measures, to compensate for loss of coral reef due to coastal construction, dredging or ship groundings.

- To improve the aesthetic appearance of reefs for tourism and/or mitigate the impacts of resort construction, in which case the initiative is often supported by a resort operator and may involve an element of tourist education and awareness raising.

- To repair damaged reefs that are now within marine protected areas (MPAs) but show few signs of natural recovery (e.g. extensive rubble areas caused by blast-fishing).

- To raise public awareness of reef resources and their conservation.

- To assist recovery of reefs impacted by disturbances such as tropical cyclones, Crown-of-thorns starfish (*Acanthaster planci*) outbreaks, or bleaching events.

Given that reef restoration is a relatively new approach, much of the work currently underway is still of an experimental nature and the primary objective is often therefore to test new techniques or improve existing ones. There may also be secondary objectives. In particular, in a number of cases, reef restoration has been used as a means of raising awareness about the importance of reefs to local economies, the threats to them and the need for their conservation.

The case studies involved two types of restoration:

Physical restoration – repairing the structure of the reef and enhancing the condition of the substrate to encourage natural recovery. Case studies 1 and 10 exemplify two different approaches to the stabilisation of substrate damaged by blast-fishing. The first, in Indonesia, involves the addition of limestone blocks to provide stable substrate which facilitated natural recovery. The second, in the Philippines, involves the use of plastic mesh to stabilise patches of rubble. Both are in relatively low-energy environments (i.e. not on the reef crest or on very exposed shores) which allowed stabilisation to be achieved at moderate cost. Case studies 4 and 9 use artificial reef (AR) structures made of concrete to provide stable surfaces for coral recruitment or transplants in areas where the reef had suffered severe storm damage. In the first case in Thailand, natural recruitment led to almost complete coral coverage and an aesthetically pleasing, *Porites* dominated reefscape within 12–15 years. In the second in Mexico, it is notable that natural recruits outnumbered transplants within 6 months, but it is too early to assess outcomes.

Physical restoration has commonly only been attempted at ship-grounding sites. Costs have ranged from US\$ 5.5 million/ha (M/V *Elpis*), through US\$ ~48.8 million/ha (M/V *Wellwood*), to >US\$ 100 million/ha (R/V *Columbus Iselin*)[2], if total compensation payments are simply extrapolated based on original areas damaged. These values ignore potential economies of scale and that, in some cases, part of the funding was for compensatory restoration, grounding prevention and other activities not directly related to restoring the damage. Nevertheless, they are indicative of the order of magnitude of costs of rehabilitation of physically impacted reef at shallow, relatively high energy sites. It is encouraging that for less exposed sites (Case studies 1, 4 and 10) relatively low-cost options can deliver varying degrees of success.

Biological restoration – the transplantation of whole colonies or fragments of coral (often "corals of opportunity", that is, detached coral fragments (or colonies) that are unlikely to survive unless rescued) was used in eight of the case studies, whereas case studies 1 and 4, having provided stable substrate, relied wholly on natural recovery processes to generate recovery. In two cases whole colonies from the reef were utilised, at least in part; in one case because large colonies were needed to create patches on a sand substrate (Case study 2) and in the other because colonies were threatened by resort construction and needed to be rescued (Case study 3). In six cases coral fragments were used and in three of these some of the fragments were reared in *in-situ* coral nurseries (Case study 3: re-use of colonies reared from fragments on coral frames as a source of further fragments, 5: all colonies transplanted were reared from fragments, and 8: some transplants were obtained from a coral farm). Case study 7 uses the novel approach of deploying arrays of coral settlement devices (CSDs) at around the time of mass spawning and then deploying those with surviving 1–2 year old juvenile corals to degraded reef areas.

The increasing use of nursery reared (farmed) corals and focus on using corals of opportunity as the primary source of transplants where feasible (Case studies 3, 6, 8, 9 and 10), rather than fragmenting healthy attached donor colonies, is a positive trend which minimises collateral damage to the reef.

An Acropora tenuis colony, three years after settling on a specially designed coral settlement device (CSD), which was then transplanted to a degraded reef in Sekisei Lagoon, Japan (S. Fujiwara).

8.3 Lessons learnt

The case studies are located at a range of very different sites in both developed and developing countries and have diverse objectives. Also the sample size is small, thus few generalities emerge and each case study should largely be considered on its own merits. However, it is instructive to compare the spatial scales and costs of the studies. Following this, we highlight a few points to consider when reading the case studies and a few general lessons. Specific lessons from individual case studies are referred to in previous chapters.

Spatial scale and costs per hectare

Most of the restoration projects involve generally small areas of reef: a few hundreds to thousands of square metres (Table 8.1). The case study involving the largest area is the attempt to rehabilitate the 2.75 ha site of the M/V *Fortuna Reefer* grounding in Puerto Rico, whereas the smallest was an experimental study involving the restoration of just 80 m^2 of reef near Eilat. Projects that are experimental in nature generally cover smaller areas (100s m^2) than those where practical restoration is being attempted (1000s m^2 to a few hectares). This is probably due to the greater funding allocated to the latter, often provided by industry or insurance companies (as compensation) or by local governments. Although it is now feasible using nurseries (Chapter 4) to generate enough transplants to rehabilitate hectares, the longer-term (>5 years) success of large-scale transplantation remains to be demonstrated in areas that need it (although it has been demonstrated in areas where neighbouring untreated reefs have recovered naturally).

Table 8.1 Approximate areas of sites being rehabilitated in the 10 case studies presented here and three of those from the *Reef Restoration Concepts and Guidelines*[1] (RRG1–5) and total costs reported. This allows the typical order of magnitude of costs per hectare to be estimated (e.g. US$10,000s or $100,000s per hectare).

Case study	Area (hectares)	Cost US$	Cost per ha US$ 000
1	0.60	40,000	67
2	0.185[†]	105,400	570
3	0.125	~62,500	500
4	0.128	~72,400	566
5	0.008	?	?
6	2.75[§]	~1.25 million	455
7	0.075	?	?
8	0.20	12,000	60
9	~0.15[‡]	25,000	167
10	0.05	35,000	700
RRG1	0.72	350,000	486
RRG2	0.06[††]	97,000	1617
RRG3	0.2 ha	62,000	310

[†] 0.016 ha of patches (4) spaced at 30 m.
[§] Area of ship-grounding impact.
[‡] Based on deployment of Reef Balls (total plan area 0.019 ha) ~3 m apart to allow fair comparison with other methods (73 Reef Balls were actually deployed over 1.85 ha or ~15 m apart on average).
[††] Based on deployment of 600 rescued colonies at a density of one per m^2.

The data from the case studies suggest that these reef rehabilitation projects have cost from tens of thousands to over a million US dollars per hectare with the median cost just below US$ 500,000/ha. The range of values is similar to that reported in earlier studies[2–4] with most projects costing in the order of US$ 100,000s/ha. Whether all the costs incurred really contributed to the rehabilitation and the desirable spacing of patches of restoration (and thus how areal costs should be scaled up) can be argued, but even if reported costs are quartered the costs of attempting reef rehabilitation remains in the US$ 10,000–100,000/ha range. Of course, not all projects were successful but the analysis gives you an idea of what it is likely to cost to attempt to restore a hectare of reef using transplantation or substrate stabilisation techniques. These costs can be compared to those for rehabilitation of mangroves (US$ 3000–510,000/ha), seagrasses (US$ 9000–680,000/ha), and saltmarshes (US$ 2000–160,000/ha)[3]. The key lesson to be learned is that the cost of *active* rehabilitation of coastal habitats is substantial and likely to be far more than the costs of implementing effective protection of the habitat that may in time allow natural recovery.

Triangular prisms of concrete pipes after 12 years of natural coral settlement and growth at Maiton Island, Thailand (N. Thongtham).

Long-term impact of restoration

Most of the restoration projects were initiated within the last decade, which means that it is still not possible to determine the long-term impact of the activities undertaken. However, case studies 1, 4 and 6 started in 1998, 1997 and 1994 respectively, and their long-term impacts can be evaluated. Case study 1 (Indonesia) shows that unstable rubble areas created by blast-fishing can be effectively stabilised in areas with moderate current using limestone boulders for a cost of about US$ 5/m² (with benefits of scale this might equate to US$ 25,000/ha). However, the authors and colleagues have now examined whether scarce funding would be better spent trying to restore already damaged areas or invested in marine patrols to enforce bans on blast-fishing[5]. Based on an economic analysis, they conclude that in this case marine protected area managers should prioritise investment in achieving compliance with regulations rather than rehabilitation.

Case study 4 (Thailand) was prompted by the destruction of an area of *Acropora* thickets by a storm and the subsequent non-recovery during 8 years of the sandy substrate that remained. Triangular prism modules made of concrete pipe were deployed to see if the coral community would return naturally, once bare stable substrate was available. Within 12 years most of the modules were more or less completely covered by live coral and resembled natural reef patches. In terms of rehabilitation the project is a marked success but in terms of restoration (see Chapter 1 for definitions) the original *Acropora* dominated thickets have been replaced by a *Porites* dominated community, thus restoration (in the strict sense) has not been achieved. This study clearly demonstrates the power of *passive* restoration at well-managed sites where local anthropogenic impacts are minimal.

Case study 6 (Puerto Rico) was an attempt to rescue and re-attach almost 2000 fragments of Elkhorn coral (*Acropora palmata* – a species listed on the US Endangered Species Act) that had been broken off by a ship-grounding in 1997. Careful monitoring over 10 years has allowed many valuable lessons to be learnt from this study and shows the formidable challenges of trying to restore sites subject to high wave exposure. Detachment due to waves, overgrowth by the sponge (*Cliona*), predation by snails (*Coralliophila*) and White Band disease were the main factors which resulted in only about 6% of the fragments remaining alive by 2008. However, without intervention survival is likely to have been even lower. Unfortunately, studies of the survival of natural Elkhorn coral fragments in a similar situation over 10 years are lacking so one cannot fully assess the costs and benefits.

Setting objectives

The importance of setting clear aims and objectives and agreeing these among stakeholders is discussed in Chapter 2. In section 2.4, the point is made that without clear objectives, it is not possible to evaluate success and it is difficult to learn lessons. To assist this, aims need to be realistic, objectively verifiable and time-bound and monitoring needs to be built into the rehabilitation plan to allow stakeholders to evaluate progress.

In reading the case study objectives (which are necessarily condensed), it is a useful exercise to consider how you would refine the objectives so that you would be able to derive quantitative and time-bound criteria for evaluating the success or otherwise of each project. Are the objectives clear? Are they feasible? What information would you need in order to define criteria for evaluating success? What monitoring would you need to discover whether targets had been met?

Social aspects

Effective stakeholder consultation and involvement is generally considered as essential to the sustainability (long-term success) of any reef rehabilitation project

(Chapter 2). The importance of this will be greater in projects where local communities are intimately involved and particularly where there are several groups of stakeholders with differing aspirations. A few of the case studies are essentially experiments by scientists and thus social impacts were incidental, generally involving opportunities to increase public awareness of reef conservation issues. However, in others (e.g. Case study 2, 8, 9 and 10), there are stated social objectives. To assess whether these were achieved, you would need to carry out some form of social or economic monitoring (e.g. via questionnaires) and also define some objective criteria or benchmarks for evaluating whether social objectives were attained. This is not easy and may require substantial resources if done properly. Although biological monitoring is routinely considered, surveys to assess socio-economic impacts of reef rehabilitation projects appear rarely to form part of the monitoring strategy. Such information would greatly assist benefit-cost analysis of past and proposed reef rehabilitation projects[3-4].

Transporting corals

In general we have recommended that where transplantation is required, the time taken to transport corals should be as short as possible and corals should be kept immersed in fresh, well-oxygenated seawater which should not be allowed to warm above ambient sea temperatures (Chapters 4–6). However, in two cases studies reported here (Case study 3 and 8) and one from the *Reef Restoration Concepts and Guidelines*[1] (RRG5) in Maldives, Fiji and New Caledonia respectively, corals were transported for up to 30–60 minutes exposed to air but shaded and either sprayed with seawater at intervals or covered in damp towels. In no cases did the corals appear to suffer significant stress, confirming earlier observations[6] that, for up to an hour, as long as corals are shaded, kept damp and not allowed to heat up, survival is not significantly worse than corals kept immersed in seawater.

Attaching corals to the substrate

At all case study sites, some attempt was made to ensure that corals remained in place once transplanted. At some lower energy sites, large transplants were embedded in the sand or placed on rubble and stabilised with rocks, and smaller transplants were wedged in holes and crevices in the reef. However, in most cases transplants were attached more securely using cement, epoxy adhesive, cable-ties, or stainless steel/copper-nickel alloy (Monel 400) wire. Wire has been found to be effective for attachment[7] and can be rapidly overgrown by coral in relatively low energy sites. However, if there is much wave energy it is difficult to prevent movement of the coral and this will result in the wire abrading the coral and the coral failing to self-attach to the substrate. In such a situation the wire will eventually corrode or break leading to detachment and loss of the transplant (e.g. 25% of fragments lost over 3 years in Case study 6).

For this reason, the authors of Case study 6 do not recommend using wire alone in high energy situations, although they note that wire may be useful to temporarily hold coral transplants in place until cement or epoxy hardens. Plastic cable-ties were also found to loosen quickly in heavy surge (Case study 6) although they were used successfully at less exposed sites (Case study 3 and 10).

A Reef Ball[TM] off Cozumel, Mexico that has been colonised by macro-algae (M. Millet Encalada).

Maintenance of rehabilitation sites

Where natural recovery processes were wholly or primarily relied on to repopulate the rehabilitation sites with corals, little or no maintenance was generally necessary (Case study 1, 4 and 10). By contrast, where transplantation of corals was carried out, some maintenance of the transplanted patches was normally required. Transplants were liable to predation by fish (e.g. butterflyfish, parrotfish) and invertebrates (e.g. the snails *Drupella* and *Coralliophila*; the echinoderms *Acanthaster* and *Culcita*) or overgrowth by macroalgae, filamentous algae, blue-green algae or sponges (e.g. *Cliona*). Maintenance activities reported involved removal of coral eating snails (*Drupella*), macroalgae/ seaweed, Crown-of-thorns starfish (*Acanthaster*), cushion stars (*Culcita*) and human rubbish, and stabilisation or reattachment of transplants that had been moved by waves or otherwise become dislodged. Coral transplants mostly appeared to recover from initial fish predation (Case study 5). The case studies suggest that a degree of maintenance of transplants to improve survival, at least during the early months, seems to be widely recognised as worth the effort involved. We are not aware of rigorous comparisons of maintained and non-maintained transplant sites to assess the cost-effectiveness of maintenance activities, but given the vulnerability of patches of coral transplants (Chapter 3), the investment is generally likely to be worthwhile.

Monitoring

Monitoring is essential (i) to evaluate the success or otherwise of reef rehabilitation projects, (ii) to identify emerging threats, (iii) to allow adaptive management and mitigate risks, (iv) to provide feedback to stakeholders, and (v) to allow lessons to be learnt (Chapters 2 and 3). Monitoring can be notionally divided into (a) regular and fairly frequent visual inspection with the aim of checking if there are any problems which require maintenance or adaptive management, and (2) less frequent systematic surveys that will allow progress towards objectives (success) to be evaluated and communicated to stakeholders. For case studies involving coral transplantation, monitoring was carried out at varying intervals, ranging from monthly to every 3–4 months during the first year and subsequently at intervals of 6–12 months. For projects relying on natural recovery processes, monitoring was carried out at intervals ranging from one to several years. Thus monitoring frequency matched the expected rate of occurrence of interesting changes to the transplanted or natural communities. For transplantation, there is an expectation that if things go wrong they are likely to go wrong early on, when the transplanted corals are most likely stressed and have not yet had time to self-attach to the substrate. Thus initial surveys tend to be at 1, 2 or 3 months after transplantation. These can provide valuable data on causes of mortality (e.g. predation, disease, detachment). Once transplants are established (after 1–2 years), then survey intervals are reduced to every 6–12 months. The wider apart the surveys, the less chance of establishing causes of mortality if corals die.

In two projects, there appeared to be initial success with low mortality levels, but high levels of mortality appeared between the 6 and 9 month surveys (Case study 8) and 12 and 15 month surveys (Case study 2). This emphasises that good, short-term (<2 years) results are unreliable indicators of longer-term success. The timescale for natural recovery of reefs from major coral loss, such as the mass-bleaching and mortality experienced in the Indian Ocean in 1998, appears to be at least 10 years; thus a realistic period over which to evaluate the success of a rehabilitation project is likely to be about 5–10 years, given that some acceleration of recovery should be achieved.

Local snorkelers laying out a patch of Acropora spp. *transplants at ~4 m depth near Fongafale, Tuvalu (D. Fisk).*

Personnel and equipment

All the case studies benefitted from varying degrees of input from experienced coral reef biologists. Although one of the roles of this manual is to reduce the need for this scientific input by providing detailed guidance to managers, until such time as reef rehabilitation projects are routinely successful (most definitely not the case now), some input from experienced biologists remains essential.

Reef restoration work often necessitates long periods underwater and is thus preferably undertaken using SCUBA gear, even in shallow depths (3–5 metres). However, where the need is for a low cost method that can be used by local people, snorkelling and free diving can be used in certain circumstances. Examples include the case studies from Tuvalu and Fiji (Case study 2 and 8), where there was limited access to diving equipment or no means of ensuring diver safety, and where the participation of the local communities was central to the projects. Luckily, in both cases depths were shallow (3–5 m and 1–4 m depth respectively) and the lagoonal areas relatively sheltered such that all or most transplants could be planted in sand, wedged in crevices or held in place by rocks. Unfortunately, for various reasons neither project was particularly successful in terms of ecological outcomes, however this does not mean that low-cost community-based approaches using free diving cannot work[8].

Acknowledgements

We wish to thank Andrew Bruckner, Robin Bruckner, Hansa Chansang, Dave Fisk, Helen Fox, Shuichi Fujiwara, Cedric Guignard, Kelly Haisfield, Yael Horoszowski, Zaidy Khan, Thomas Le Berre, Marinés Millet Encalada, Laurie Raymundo and Nalinee Thongtham for sharing the details of their reef rehabilitation projects with us. We hope you find them instructive.

References

1. Edwards, A.J. and Gomez, E.D. (2007) *Reef Restoration Concepts and Guidelines: making sensible management choices in the face of uncertainty.* Coral Reef Targeted Research & Capacity Building for Management Programme, St Lucia, Australia. iv + 38 pp. ISBN 978-1-921317-00-2 [Download at: www.gefcoral.org/Portals/25/workgroups/ rr_guidelines/rrg_fullguide.pdf]

2. Spurgeon, J.P.G. and Lindahl, U. (2000) Economics of coral reef restoration. Pp. 125-136 in Cesar, H.S.J. (ed.) *Collected Essays on the Economics of Coral Reefs.* CORDIO, Sweden. ISBN 91-973959-0-0 [Download at: www.reefbase.org/download/download. aspx?type=10&docid=6307]

3. Spurgeon, J. (1998) The socio-economic costs and benefits of coastal habitat rehabilitation and creation. *Marine Pollution Bulletin,* 37, 373-382.

4. Spurgeon, J.P.G. (2001) Improving the economic effectiveness of coral reef restoration. *Bulletin of Marine Science,* 69, 1031-1045.

5. Haisfield, K.M., Fox, H.E., Yen, S., Mangubhai, S. and Mous, P.J. (2010) An ounce of prevention: cost-effectiveness of coral reef rehabilitation relative to enforcement. *Conservation Letters.* (online)

6. Harriott, V.J. and Fisk, D.A. (1995) Accelerating regeneration of hard corals: a manual for coral reef users and managers. *Great Barrier Reef Marine Park Authority Technical Memorandum,* 16, 42 pp.
[Download at: www.gbrmpa.gov.au/__data/assets/pdf_file/0017/3554/tm016_full.pdf]

7. Heeger, T. and Sotto, F. (eds) (2000) *Coral Farming: A Tool for Reef Rehabilitation and Community Ecotourism.* German Ministry of Environment (BMU), German Technical Cooperation and Tropical Ecology program (GTZ-TÖB), Philippines. 94 pp.

8. Gomez, E.D. (2009) *Community-based restoration: the Bolinao experience.* Coral Reef Targeted Research & Capacity Building for Management Program, St Lucia, Australia. 4 pp.
[Download at: www.gefcoral.org/LinkClick.aspx?fileticket=5ysYNhGL45l%3d&tabid=3260&language=en-US]

Case study 1

Substrate stabilisation to promote recovery of reefs damaged by blast fishing.

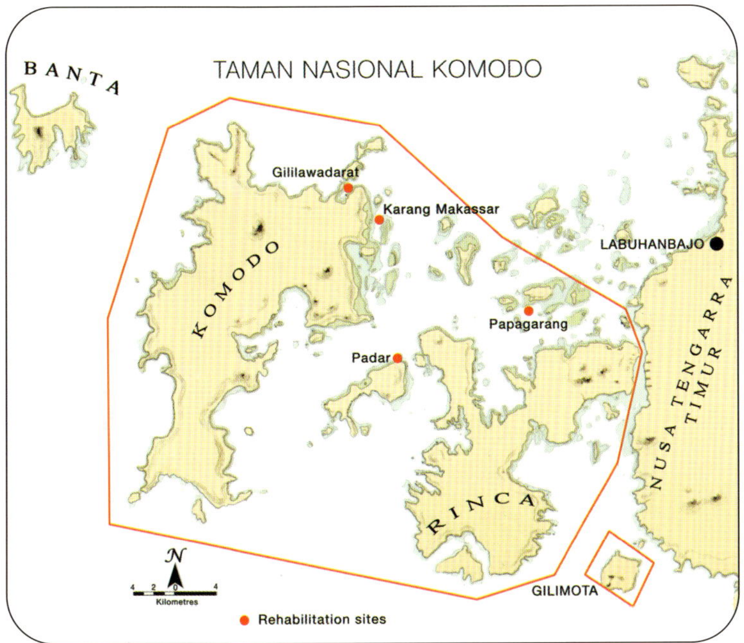

Komodo National Park (Taman Nasional Komodo), Indonesia (1998 – 2008).

Background

Since the 1950s, about half of the coral reefs in the 1,817 km^2 Komodo National Park (KNP) have been damaged by blast fishing. In 1995, park authorities initiated a patrolling programme that reduced blast-fishing by at least 80%. However, although coral larvae are plentiful and water quality good, natural recovery of heavily blasted sites did not occur. Thus, although blast fishing is now relatively rare, there are still large rubble fields of dead coral fragments that move with the current and limit natural regeneration by abrading or smothering new coral recruits, causing high juvenile mortality and inhibiting coral growth.

Objective

The aim was to increase hard coral coverage, and thus marine biodiversity, in blasted areas that previously supported coral reef communities, by stabilizing the substrate using low-cost, low-tech techniques. An additional objective was to determine the most effective and economically viable configuration of rock piles for increasing coral growth and limiting rubble encroachment.

Methods

A baseline survey was conducted to assess live coral cover, the presence of coral recruits, and current flows at different locations. In an initial pilot study, three rubble stabilization techniques were tested: netting (c. 5 cm mesh fishing net), concrete slabs, and rock piles. Although corals initially recruited using all three methods, the netting was eventually covered by rubble, the concrete slabs were frequently overturned, and rubble started filling in around the rock piles. Because the rock piles could be made larger and built up above the rubble fields, they showed the most promise and were used for the larger scale study reported here.

Four rehabilitation sites with large areas of rubble and limited (<1%) live coral cover (so that rocks could be unloaded from boats without damaging existing coral) were selected. Limestone rocks were quarried in nearby western Flores and transported by truck and boat to the rehabilitation sites. At each site, four rock piles of different designs were installed from March to September 2002 using approximately 140 m^3 of rock per installation. The rocks were thrown into the water from boats and then rearranged where necessary by divers using SCUBA at depths of 5–10 m. The four designs were:

1. Rock piles 1–2 m^3 in size spaced 2–3 m apart (covers most area per m^3 of limestone rock, but leaves the majority of the rubble unstable).

8

2. Complete coverage of the area with rock c. 75 cm high (no loose rubble within treatment area, but covers least area per m³ of limestone rock deployed).

3. "Spur and groove" rows perpendicular to the prevailing current c. 75 cm high and 2 m wide, spaced 2–3 m apart (based on naturally occurring reef formations in high wave energy locations that may enhance settlement of coral larvae by creating turbulent flow as spurs obstruct the current).

4. "Spur and groove" rows parallel to the current c. 75 cm high, 2 m wide, spaced 2–3 m apart (based on naturally occurring reef formations that may allow rubble to be flushed through the grooves).

The different designs were chosen to investigate the trade-off between more complete coverage and thus stabilization of rubble (greater cost/m² restored), versus greater total area covered, but with rubble free to move with the current in between rock piles (less cost/m² restored), for the same approximate volume of rock per installation. Results were compared to untreated rubble control plots at each site.

Monitoring

Due to the difficulty of identifying corals, trained scientists carried out the monitoring. Monitoring surveys were conducted 1, 3 and 6 years after installation using a standard protocol and collected the following data:

Diver surveying a rehabilitation site, measuring the size and taxon of coral organisms within part of a 1 x 1 metre quadrat (S. Mangubhai).

• Size and taxon (usually to genus or family) of each organism (hard corals, soft corals, sponges and other sessile organisms) present on rocks in six 1 m x 1 m quadrats in each treatment and control site.

• Size of rock piles (to measure rubble encroachment).

Control rubble sites near each rehabilitated site were surveyed to collect data on natural regeneration. Differences in fish populations were assessed through stationary video with no diver nearby, one year after installation and by UVC (underwater visual census) three years after installation.

Ecological outcomes

After 6 years, little to no natural regeneration occurred at the control, untreated blast sites with hard coral cover remaining at <1%, but even at the least successful treatment sites, live coral cover increased significantly. Fish aggregated around the rock piles almost immediately after installation. After 6 years, hard coral coverage of the rocks was as low as 8% at the least successful site (complete coverage at Gililawadarat), and as high as 43% at the most successful site (parallel rows at Papagarang). There was high variability between rock pile configurations and between sites, with no clear best configuration option. The one constant was the limited coral growth at Gililawadarat, a low current site, for all rock pile configurations. There was high variability between locations in terms of rubble encroachment: rubble filled in at high current sites (around piles and between grooves of perpendicular and parallel rows, and on top of complete coverage sites), and there was sedimentation at the low current sites. Greatest success was achieved at sites with moderate levels of current.

New coral heads and fish aggregating near a rehabilitated rock pile. Note rubble in background (H. Fox).

Social aspects

The project was discussed with local communities, park rangers, tourism operators, and presented to Indonesian reporters. Involving the community and park rangers can create a sense of responsibility for managing and protecting coral reef resources and educate people about the importance of healthy reefs. Local people benefited from the salaries paid, and the increased diversity will have increased the tourism value of the area.

Resources required to stabilize c. 6000 m² of rubble

Human resources: The rocks were deployed using a cargo boat with 8–12 crew to load, transport and unload rocks; 1 boat driver and 1 volunteer, coordinated by a team of 2 divers (1 scientist and 1 park ranger) finalized the rock configurations underwater. Park rangers measured the rock pile sizes. Monitoring was undertaken by a single scientist/consultant (to provide continuity), partnered with another scientist trained in coral identification.

Financial resources: Total budget c. US$40,000, provided by The Nature Conservancy (TNC) and the Packard Foundation. Rehabilitation total budget: c. US$30,000. Cargo boat rental for 76 trips (US$17,000); speedboat to transport divers (fuel cost: US$3,380); 2275 m³ of rock (910 truckloads, US$7,078); park ranger stipends: US$2,500 (10 days/month x 7 months); external consultants: US$10,000. Scientist salaries covered byTNC.

Cost per m² of each design was: c. US$17 for complete coverage; c. US$5 for spur and groove rows; c. US$3 for rock piles; c. US$5 on average.

Time: 6 months for project design and implementation. 8–10 boat-days per monitoring trip.

Lessons learnt

Limestone rocks can be an effective and relatively inexpensive method for stabilizing substrate after blast fishing. This technique may be viable for marine protected areas that have easy access to rocks, and boats in which to transport them, provided that the blasting is halted and that coral larvae are abundant. In the case of KNP, the MPA has a 25 year management plan that is relatively effectively implemented.

In this study, rocks in rows mimicking spur and grooves parallel to the current developed highest average hard coral cover after six years, but with high variability between sites. More studies are needed to determine the best rock configurations for differing current/depth conditions.

Some tabulate corals became victims of their own success, falling off rock piles that could no longer support their weight. Perhaps some type of cement to strengthen piles could eliminate this problem, although this would complicate installation.

Recently-installed parallel rows of limestone rocks (H. Fox).

Rock stabilisation (cost of materials, transportation, boat rental, and labour totalling c. US$5/m²) was an inexpensive method compared with some other techniques in the literature. Costs could be further reduced (potentially by 50% or more) if stabilisation were to be undertaken at a larger scale by, for example, negotiating better rates or having a boat built and a crew hired specifically for the project. However, it should be noted that costs are considerably cheaper in Indonesia than in other parts of the world, and projects in low-lying atolls, such as the Maldives, would not have access to rock quarries.

Reference

Fox, H., Mous, P., Pet, J., Muljadi, A. and Caldwell, R. (2005) Experimental assessment of coral reef rehabilitation following blast fishing. *Conservation Biology*, 19(1), 98-107.

Case study provided by: **Helen Fox** and **Kelly Haisfield**, Conservation Science Program, World Wildlife Fund - US, 1250 24th Street NW, Washington, DC 20037 – 1132, USA; Tel: +1 (202) 495-4793; Fax: +1 (202) 293-9211; E-mail: Helen.Fox@WWFUS.org, Kelly.Haisfield@WWFUS.org

8

Case study 2

Transplantation of coral colonies to create new patch reefs on Funafuti Atoll, Tuvalu.

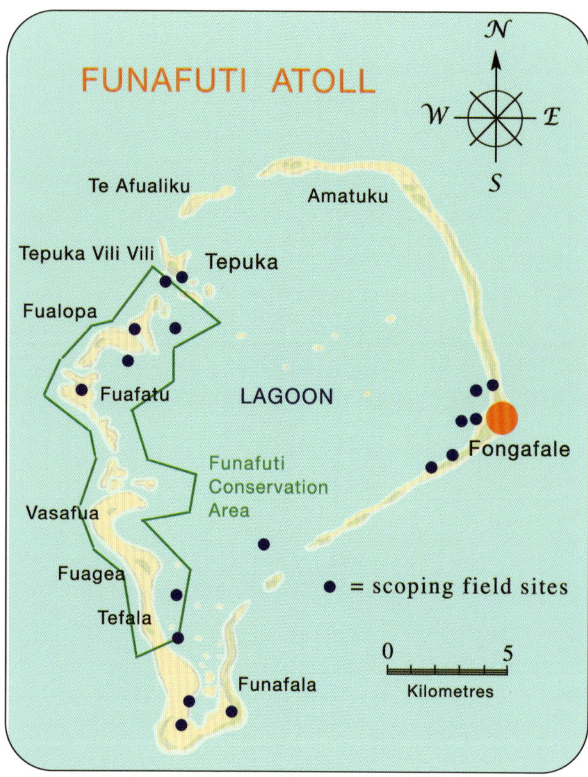

Background

This collaborative project between the Foundation of the Peoples of the South Pacific International (FSPI) and Tuvalu Association of NGOs (TANGO) was initiated, as part of the Coral Reef InitiativeS for the Pacific programme (CRISP), as a result of local concerns about the decline in fish catches in the lagoon, considered to be due to loss of extensive branching coral thickets caused by macro-algal overgrowth and predation by the corallivorous snail *Drupella cornus*. Poor groundwater quality and nutrient input due to bad land management could have led to the high macro-algal cover; the apparent absence of algal grazers (no urchins were observed and only low numbers of fish grazers) could also have been a factor. A previous UNDP-GEF International Waters Programme project found nutrient contamination of groundwater on the adjacent populated cay, but the extent of leakage of this into the lagoon is unknown.

Objectives

The ecological aim was to create suitable habitat for juvenile fish in an area that is currently low in fish and to recreate branching coral thickets on sandy substrate that was devoid of harmful macro-algae and *Drupella*. The intention was that the patches would be substantially larger than the surviving adjacent reference patches, would resist wave action, and would be structured with a live-coral canopy and a dead-coral understorey. The project also assessed the cost-benefits of engaging a local NGO, fishers, school children and others in the local community to carry out low-tech reef restoration efforts. There were also social objectives: to raise awareness of the importance of healthy reef habitats for sustainable fisheries and demonstrate that current disturbance factors on Funafuti (poor water quality, high abundances of territorial damselfish (*Stegastes* spp.), macro-algal overgrowth, corallivorous gastropod infestations) are major contributors to poor coral cover which in turn influences the fringing reef fisheries' biomass.

Methods

A three-day scoping study of the inner lagoon edge was undertaken, covering most of the habitat adjacent to the populated cays, and several sites within the distant and unpopulated 33 km² Funafuti Conservation Area (FCA) in the western lagoon (see map). The results of the scoping survey were reported to stakeholders and used as a basis to select the best restoration method and the translocation site. The FCA was too remote from human populations to offer many of the main objectives although it was devoid of the major disturbances found on other reefs. The study site was chosen because of its proximity to a nearby school and the anticipated awareness raising value, its high visibility (which was thought to afford some measure of protection from poaching or intrusions), and its ease of access for fisheries officers involved in both the translocation and the ongoing monitoring. The site is about 165 m from the lagoon side beach, and consists of a sandy substrate habitat adjacent to the fringing reef slope, at 4–5 m depth, and at least 30 m from the fringing reef slope and other

Translocation of coral transplants from donor site to transplant site (D. Fisk).

scattered patch reefs on the sand. Although nearby fringing reefs had high levels of macro-algae cover and *Drupella* densities, these were not present at the selected site, although periodically, rubbish was deposited in the lagoon as a result of flooding from storms and spring tides.

Five free-divers collected about 240 colonies of 2–3 species of branching *Acropora* (predominantly *A. intermedia* and *A. muricata)* by hand from the fringing lagoon reef in November 2006. One or two large colonies were collected at a time by each diver and cleaned of macro-algae and *Drupella* before being taken to the transplant site by swimming. The colonies were never lifted above water, and time from donor site to transplant site averaged 5 minutes; the live coral portions of the colonies were handled as little as possible. The colonies were placed at 3–5 m depth on the sandy substrate in an upright position, with the dead basal portion buried in sand. Four 28–50 m² patches of transplants were established 30–40m apart. The average patch size was 41 m² and each contained about 60 colonies, placed about 1 m apart.

Maintenance of the new patches was undertaken during each monitoring session and involved removing *Drupella* and macro-algae, and re-establishing colonies in the upright position if they had been moved or dislodged.

Monitoring

Monitoring took place over 15 months, at 0 (baseline), 2, 4, 8, 9, 12 and 15 months after translocation, and was carried out by the restoration team and local villagers. Each monitoring survey included:

- Coral health: indicated by the volume of live and dead coral in each patch, incidence of bleaching or disease, and stability of transplants.

- Disturbance: indicated by presence of *Drupella*, Crown-of-thorns starfish (*Acanthaster planci*), rubbish and macro-algal overgrowth.

- Growth of suitable fish habitat: indicated by the volume of live coral canopy and associated dead understorey corals within a patch, and the length and breadth of each patch.

- Fish and invertebrate presence and recruitment: number, size, family and/or species, found within 1 m of each

patch were recorded. All fish in each patch were counted, and fish families and species were categorized according to major feeding groups and were assigned to one of five size-classes.

Ecological outcomes

After 15 months, the visual census showed an average mortality rate of 74% of the transplants, and was as high as 95% at one patch. The mean volume of live coral decreased from 82% initially (18% of initial patch volume was dead coral) to 55% after the first 2 months and 39% after 4 months. Subsequently the volume of live coral increased to 50% after 8 months, 59% after 9 months, and 61% after 12 months. However, at 15 months, mean volume of live coral dropped dramatically to 26%.

The cause of the major mortality at 15 months was not clear. Throughout the monitoring period, the transplants, as well as *Acropora* colonies on the adjacent fringing reef and other fringing reefs along Funafuti's eastern coast, commonly showed a white band at the base of many branches, which was most likely the result of *Drupella* predation or disease (e.g. white syndrome). There were no physical conditions during the project to cause serious bleaching, and the partial bleaching that was observed may have been due to smothering by or contact with sand during stormy periods, and thus more related to abrasion than thermal stress. In February 2007, four months after translocation, the patches experienced storms which coincided with high tides, resulting in stronger than normal waves and currents in the lagoon and the deposition of rubbish from the adjoining cays. There were no other natural disturbances, although by 8 months, there had been major sediment reworking by burrowing shrimps causing dense concentrations of cone shaped sediment mounds, which resulted in smothering of some transplanted colonies.

There was a gradual build up in both the number and diversity of fish species at the patches. Many of the fish

8

Local skin diver monitoring a transplanted patch (D. Fisk).

were juveniles of species targeted by local fisheries. After 15 months, a total of 85 fish were counted on the four patches, most of which were damselfish (39%), goatfish (28%) and wrasses (21%). About half the species were carnivores and half omnivores. Most fish were small in size or juveniles (1-10 cm, with most less than 5 cm).

Social aspects

Community and government members (Ministers for Fisheries, Environment, and Community Affairs, TANGO members) were addressed both before and after the initial scoping exercise, when the results of the survey of the status of the lagoon reef habitats were presented, and questions answered. Talks were given to school children and they were taken on conducted snorkel swims over the patches in order to raise awareness. The local school was also made responsible for 'looking after' the transplanted patches. At the end of the project, awareness raising was conducted on the causes of reef degradation and the importance of coral reefs to the community. Knowledge of reefs and source of local impacts was considered to have increased.

Resources required to establish, maintain and monitor c. 160 m² of coral patches

Human resources: 9 people were involved in total including: 2 FSPI staff (project logistics, coordination and participation in transplantation); 2 scientific consultants (advice, design, report production, project management); 4 local fisheries staff (transplantation and monitoring); 1 TANGO staff (monitoring, community aspects/liaison). The scientific consultants provided initial training for the fisheries, TANGO and FSPI staff.

Financial resources: Overall FSPI budget c. US$105,400. Salaries: scientist – c. US$12,700; FSPI staff – c. US$14,000; TANGO – c. US$7000. Travel: local c. US$2100; regional c. US$5600. Equipment: c. US$1400. Living expenses: local DSA c. US$770; regional DSA: c. US$2100.

Time: Project design: 6 days including 4 days scoping field work. Collection, transportation, and transplantation of corals = 2 days. Monitoring = 1 day per survey.

Lessons learnt

Season of transplantation: The live coral mortality in the first 4 months indicates that the corals were initially stressed by transplantation, perhaps because they were moved to water 2–3 m deeper than where they were collected, or because they were moved at a time of the year (November) when sea temperatures are rising and close to the annual maximum causing corals to be under more stress. Stress due to the latter could have been avoided if transplantation had taken place in June–August (see recommendation in

Edwards and Gomez, 2007: p. 23), but project delays and timing of implementation meant that this risk was unavoidable.

Site selection: The transplants appeared to be affected by rubbish accumulation and possibly by poor water quality associated with heavy rain that drained waste water from the adjacent populated island. This underlines the importance of choosing a transplantation site that is free of adverse human impacts. However, there is a trade-off between this and other factors (e.g. ease of maintenance and monitoring). In this case, the site choice was a compromise between ideal environmental conditions and suitability for the low-cost/low-tech/community-based approach, which required easy access from shore to allow visits by local people including school groups.

Maintenance: If a sub-optimal site location is unavoidable, there needs to be frequent maintenance, in this case to remove rubbish and predators such as *Drupella*, especially at times of the year when these threats are more critical (e.g. after storms, heavy rain, and during the summer months).

Disturbances: The physical disturbance of the transplanted colonies that occurred during storms could perhaps have been reduced by placing transplants closer together in each patch (to give each other mutual support as in mature branching coral thickets) or having a series of small interconnected patches no more than 1 m apart. The unpredictable occurrence of a large settlement of burrowing organisms that interfered with the transplants underlies the uncertainties of choosing any site for restoration.

Local investment in surveillance and monitoring is necessary to keep costs low. Education is needed to ensure raising of awareness. In this project, greater involvement and participation could have been sought from the broader local community at the beginning as well as the end of the project so as to maximize the awareness value.

References

Fisk, D.A. (2007) Funafuti Atoll coral reef restoration project (Republic of Tuvalu) – Baseline report. CRISP, June 2007. 42 pp.

Fisk, D.A. and Job, S. (2008) Funafuti Atoll (Republic of Tuvalu) – Coral reef restoration project, 1-3-6-9 and 15 months post-trial – Monitoring report. CRISP, July 2008. 34 pp.

Reports available on: www.crisponline.net/CRISPProducts/Reefrestoration/tabid/310/Default.aspx

Case study provided by: **Dave Fisk** and **Zaidy Khan**, FSPI (Foundation of the Peoples of the South Pacific International), P.O. Box 18006, Level 2, Office 2, Victoria Corner Building, Suva, Fiji; Tel: +679 331 2250; Fax: +679 331 2298; E-mail: dave.fisk@gmail.com, zaidy.khan@gmail.com.

Case study 3

Transplantation of coral fragments and colonies at tourist resorts using coated metal frames as a substrate.

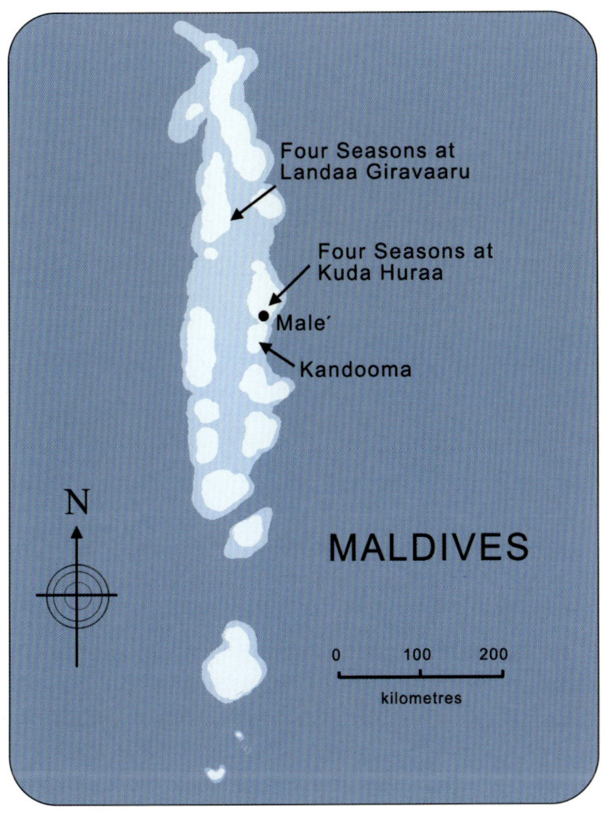

Maldives: Four Seasons resorts at Landaa Giravaaru (2005) and at Kuda Huraa (2007); and Kandooma resort (2008). The project is ongoing.

Background

In 1998 an El Niño Southern Oscillation associated warming event caused 90% shallow water coral mortality in the Maldives following mass-bleaching. Branching acroporids and pocilloporids were most affected, while the massive and encrusting forms (poritids and faviids) were less impacted. Tourism is the mainstay of the Maldivian economy, and the rate of resort development has increased dramatically in the recent past, together with coastal infrastructure construction necessary for the economic development of the country. Direct physical damage to coral during construction/dredging is the main human impact on the coral reefs. It is not at present compulsory to salvage corals impacted by coastal construction works.

To mitigate for impacts of construction, some resort operators have funded coral transplantation activities. An early attempt involved coral transplantation to Reef Balls™ at the Four Seasons resort at Kuda Huraa, however, factors such as cost, difficulty of deployment, low survival of transplants due to sedimentation and predation by cushion stars (*Culcita*) led to the search for another technique. In 2005, Four Seasons Resorts Maldives contracted a local consultancy company, Seamarc, to develop a new technique to mitigate for coral loss from construction of jetties, water villas and a supply channel at the new resort at Landaa Giravaaru. The method involves attaching coral fragments to coated metal frames of varying design (called 'coral frames'). The preliminary successful survival and growth led to replication of the experiment at two other resorts in 2007–2008.

Objective

The main objective of this project was to salvage as many coral colonies as possible during resort development using both fragmentation and transplantation of non-massive forms and translocation of the biggest massive coral colonies to improve the aesthetics around resort structures and provide easily accessible and sheltered snorkelling areas for tourists.

Methods

The method employed relies on locally made coral frames. These are made of rebar which is welded into the desired shapes and then has two coats of polyester resin applied to prevent rusting. Beach sand is incorporated into the coating to increase surface roughness and enable better attachment of corals as well as to improve the aesthetics. These structures allow corals to be quickly and easily attached with cable-ties, raise corals above the seabed (reducing sedimentation and abrasion in areas of sand and rubble, and predation by corallivores such as the cushion star, *Culcita schmedeliana*), and are light enough to be easily deployed. Several frames can be combined to create a "reefscape" on either loose or hard substrates. The coral frames act like *in-situ* nurseries for the attached fragments.

Transplantation sites were selected by surveying the reefs at each resort to assess their suitability, in terms of risk from a) sedimentation, b) waste disposal and c) boat movements, as well as criteria such as aesthetics and accessibility. For the first project in 2005, coral fragments and colonies were collected from the reef flat of Landaa Giravaaru, Baa Atoll and transported by boat, covered in a wet towel to protect them from the sun, for about 30 minutes to the transplantation site. For subsequent projects, only broken coral fragments (which would normally have a low chance of survival) found lying on the seabed were collected.

Coral fragments (in total c. 50,000) and whole colonies are attached to coral frames using cable ties. The first frames were flat and ~4m² in area, but later it was found that 0.6–2 m diameter dome-shaped ones were easier and less time-consuming to build, and more resistant to environmental stresses. However, flat fames are still used for very shallow areas. The frames were placed at 1–10 m depth on stable pavement or sandy areas. Initially, 500 m² of frames were transplanted in 2005 at Landaa Giravaaru, using both fragments and entire colonies of over 25 species, 80% of which were acroporids and pocilloporids. After one year survival was over 90%. Thus projects at Kuda Huraa and Kandooma were started in 2007–2008; these used only detached fragments found lying on the sea bed. The work was undertaken using both SCUBA and free diving.

Broken fragments found near the frames (presumably detached from the transplants) and fragments pruned from transplants that are growing too close together are harvested by hand or using a chisel, and attached to new frames using cable-ties to generate second generation transplant material. This reduces impacts on the natural reef.

Monitoring

A marine biologist is employed at each resort to carry out monthly monitoring after training by Seamarc. Every month, photographs are taken on four sides of each coral frame, and of four fragments on each frame. Fish life, sedimentation, cover, mortality, predation and infestation are recorded visually. The percentage coral cover is estimated by the surveyor and checked on the photographs.

Ecological outcomes

After one year, survivorship at all locations was over 90%. Small fragments had a faster growth rate than entire colonies. On average, within 1 to 2 years, growth was

Dome-shaped "coral frame".
Left: *Soon after attachment of* Acropora nasuta *fragments with cable ties in November 2007 (Seamarc).* Right: *About 15 months later in March 2009 (Seamarc).*

Left: An early design with two flat frames welded one above the other, showing coral growth 18 months after transplantation of fragments (top frame) and whole colonies (bottom frame) rescued from resort development. The lower tray is 2 m square and the upper 1.5 m square. Transplanted species include Acropora muricata, A. digitifera, A. samoensis and A. hyacinthus (Seamarc). Right: The same structure 36 months after transplantation (Seamarc).

sufficient to transplant 3 times more coral frames using second generation fragments. By summer 2008, the frames covered 900 m² at Landaa Giravaaru, 300 m² at Kuda Huraa, and 50 m² at Kandooma giving a total area of frames of 1250 m², using c. 50,000 small fragments, at a density of 20–60 fragments per m² of coral frame (average 40/m²), with the density dependent on the rate of growth of the species. Fish, shrimps and molluscs were also attracted or recruited to the 'coral frames'.

Social aspects

Four local fishermen at Fulhadhoo, a small island of Baa atoll, have switched from reef fishing to making coral frames full-time (US$350/month). The project also increases public awareness of coral conservation issues among both staff and tourists at the resorts.

Resources required

Human resources: 1 part-time consultant; 4 part-time biologists (1 per site); 4 full-time local labourers; the consultant was responsible for training the biologists and labourers.

Financial resources: The overall cost of establishing transplanted coral frames is estimated at US$50–200/m²: salaries 25%, materials 50%, living expenses 25%.

Time: Transplantation of 1 m² takes 30 minutes. Monitoring takes 3–5 minutes per frame each month.

Lessons learnt

The light and inexpensive substrate of the 'coral frame' reduces predation. Cushion stars (*Culcita*) cannot climb on it and the high fragment density appears to limit impacts from parrotfish (Scaridae) grazing. The open dome structure appears to reduce sedimentation due to good water flow; this is also beneficial for coral growth. After an initial transplantation of coral from the wild, subsequent frames can be populated from pruned or detached fragments from existing frames after 1–2 years growth.

References

Guignard, C. and Le Berre, T. (2008). Staghorn corals transplantation, a method to increase axial polyps (AP) quantities of transplants using cicatrisation tissues high AP generation potential. Applications on *Acropora microphtalma*, *Acropora muricata* and *Acropora pulchra*. ICRS Oral Abstract 24-12.

Le Berre, T. and Guignard, C. (2008). The coral trays, a simple method to create large sustainable artificial reefs. ICRS Poster Abstract 24.1117.

See also Reefscapers website at: www.reefscapers.com and www.seamarc.com/CoralTrays.htm

Case study provided by: **Cedric Guignard** and **Thomas Le Berre**, Seamarc Pvt. Ltd, Fourth Floor, M. Maahuraa, Kulhidhoshu Magu, Male', Maldives; Tel: +960 33 16 26; Fax: +960 33 65 75; E-mail: thomas@seamarc.com

Case study 4

Use of artificial substrates to enhance coral and fish recruitment in Phuket, Thailand.

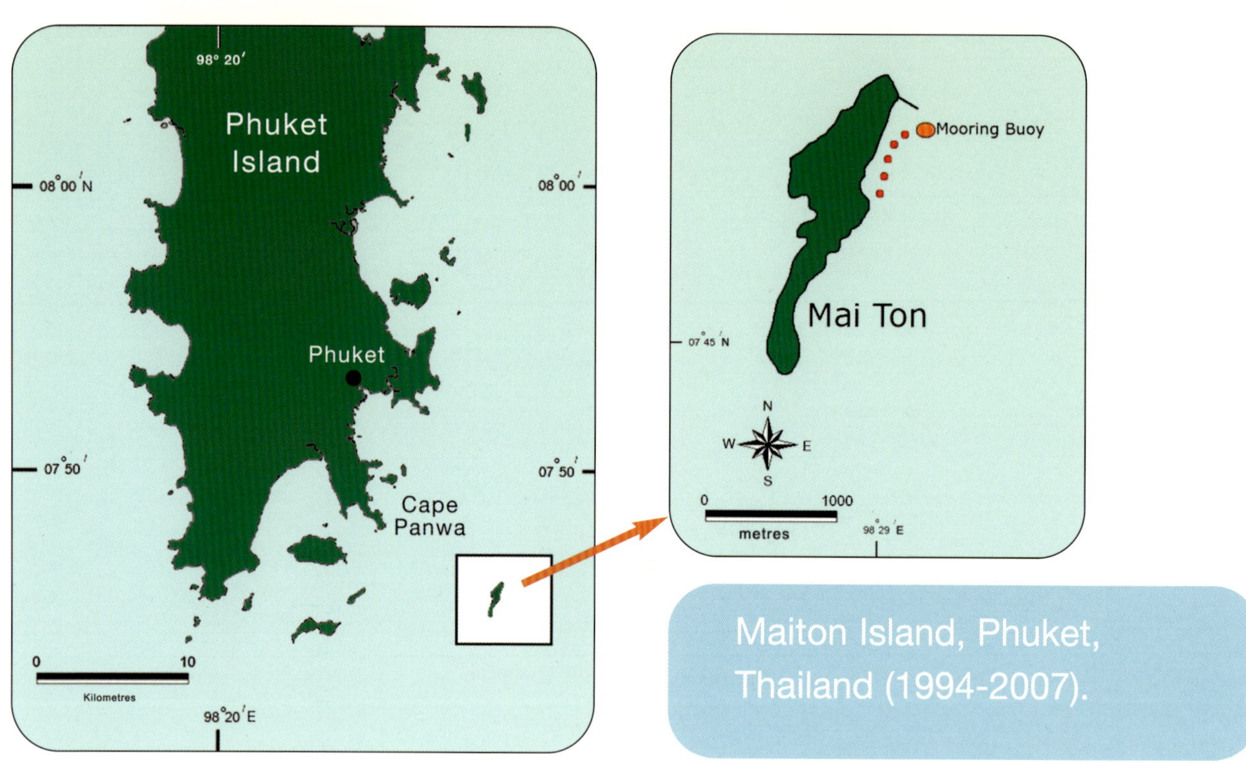

Background

In May 1986 the coral communities at Maiton Island and at Cape Panwa on the south-east of Phuket Island in Thailand were both severely damaged by a storm. Thickets of *Acropora* spp., which had dominated both sites, were destroyed and in places piles of *Acropora* rubble almost 2 m high were washed ashore. The Cape Panwa reefs recovered well with coral cover in the upper zone increasing from ~0% to 52% within 5 years. However, at Maiton Island even after 8 years there was no discernible recovery. The area formerly occupied by the *Acropora* thickets was largely sand and the surviving natural hard substrate was mainly small patches of massive *Porites* that had survived the storm (although much *Porites* was also washed ashore by the storm waves). It was unclear whether the absence of recovery at Maiton Island was due to a lack of larval supply or a lack of suitable substrate. To test this, concrete modules were deployed at the site in 1994 and have been monitored for over 12 years.

Objective

The aim of the deployment of the artificial substrates was to discover whether the site would recover naturally once substrates that offered surfaces for natural coral settlement and topographically diverse habitat for reef fish and other fauna were provided. Several different types of artificial substrate modules were tested.

Methods

Initially, 225 triangular (50 cm x 50 cm x 50 cm) concrete modules were deployed using SCUBA diving. These modules were made of 3 (low complexity), 6 (medium complexity) or 10 (high complexity) 50-cm long, concrete pipes of 20 cm, 15 cm or 10 cm internal diameters respectively, cemented together into triangular prisms. 25 modules of each type were deployed on three replicate 5 m x 5 m sandy areas at 4 m depth, covering a total area of 225 m^2 (25% of plan area being artificial substrate).

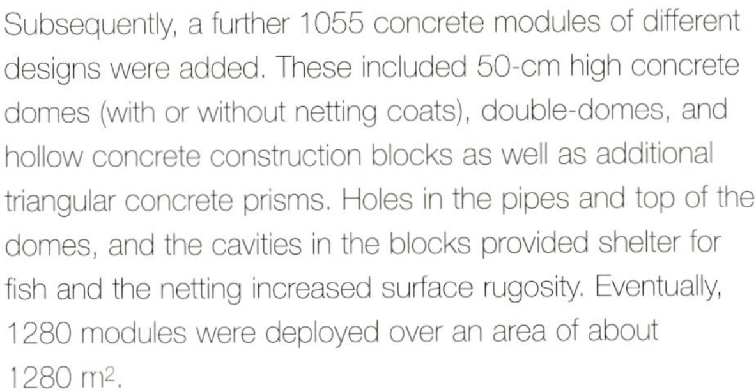

High complexity triangular prism module 4 years after deployment (N. Thongtham).

High complexity triangular prism modules 7 years after deployment (N. Thongtham).

Subsequently, a further 1055 concrete modules of different designs were added. These included 50-cm high concrete domes (with or without netting coats), double-domes, and hollow concrete construction blocks as well as additional triangular concrete prisms. Holes in the pipes and top of the domes, and the cavities in the blocks provided shelter for fish and the netting increased surface rugosity. Eventually, 1280 modules were deployed over an area of about 1280 m².

Monitoring

Colonization of a subset of 60 triangular prism modules (20 of each complexity) by sessile organisms (especially corals) and by fish has been monitored since 1994. The frequency of surveys has varied depending on the parameter being measured. Coral recruits first became visible to the naked eye 18 months after deployment and were monitored in detail at 25 and 31 months. Initially, data on the genus and size of each coral recruit that had settled on 60 of the modules and selected areas of adjacent natural reef were collected. Subsequently, coral recovery was monitored using density of species/genera and areal cover. Semi-quantitative data were obtained from photographs.

Fish were monitored at initial deployment and then at 4, 19 and 85 months afterwards. At each survey, fish were counted for 5 minutes in 5 m x 5 m quadrats, each containing 25 triangular modules of each type, and on control plots. Control quadrats were selected in adjacent areas with 20–40% live coral cover, predominantly comprised of *Porites lutea*. Counts were repeated three times for each survey.

Systematic monitoring was carried out for 7 years with subsequent visits to photograph the development of the coral community on the modules and study interesting developments. The structures added later were not monitored in detail.

Ecological outcomes

Initially, the density of coral recruits on the triangular prism modules was 20–40 times greater than that on the natural reef and the more complex substrates had significantly more coral recruits. After 7 years, 16 genera of coral had settled on the modules, the commonest being *Porites lutea*, *Millepora* sp., *Acropora* spp., and *Pocillopora damicornis*, and about 53–60% of the module surfaces were covered in live coral. There were no differences in the growth forms or species of coral that colonized among the three levels of complexity of triangular prism modules. The dome modules were less successful.

Colonization of fish in terms of number of species and individuals was rapid within the first 4 months largely due to immigration of fishes from nearby coral patches. Fish assemblages did not differ among types of modules and did not differ from communities on the natural reef nearby. Eighty-eight species in 23 genera were found, including migratory, seasonal visitors and resident species. Between 7 and 9 years, as corymbose *Acropora* colonies became established, the abundance of damselfish species that shelter in branching corals increased ten-fold, with *Pomacentrus moluccensis* and *P. adelus* reaching a density of 22–25 per 25 m² after 9 years.

Within 12 years of deployment, most triangular prism modules were more or less completely covered by live coral and resembled natural reef patches.

Social aspects

The artificial reef is managed with the help of a resort on Maiton Island, which uses it for snorkeling. The project has been well received by the diving community in Thailand and is a demonstration site for reef rehabilitation.

Resources required to establish 1280 m² of habitat

Human resources: About 15 people, comprising researchers, technicians, labourers and boat operators, conducted the deployment of the concrete modules. About 20 sport diver volunteers helped to arrange the concrete modules on the seabed according to the experimental design. For monitoring, 1–3 researchers and a speed boat crew were needed for each survey.

Financial resources: The concrete modules and their installation cost c. US$ 72,400, with local government paying for some module construction. Salaries of personnel and operational costs of monitoring were part of the regular expenditure of the Phuket Marine Biological Center on coral reef research and are additional to this figure.

Time: Project design (preliminary survey of project sites, paper work for module construction) and module construction took six months. Transportation and installation of the initial 225 modules to the site took two days, and the subsequent 1055 modules were installed in 2 weeks. Monitoring required one day for a general observation survey and 3 days for a quantitative monitoring survey.

Lessons learnt

The project is considered a success in the long term (over more than a decade) and the outcome in terms of both fish and coral community is very similar to the adjacent natural reef. However, the original *Acropora* thickets were not restored and the site remains dominated by the massive *Porites* that survived the storm. The outcome can thus be considered a successful rehabilitation but did not restore the pre-disturbance coral community. Transplantation of *Acropora* branches at the site was not considered possible due to the level of wave exposure.

No particular care was required in terms of maintenance and monitoring – natural recovery processes were just allowed to proceed. This method is recommended for damaged reefs where the physico-chemical environment is still favorable for coral growth but where stable substrates for coral settlement are lacking.

Careful site selection was critical to the success. The resort on Maiton Island looks after the site and ensures no damaging human impacts on the reefs. Initial installation required considerable manpower and this could be obtained at negligible cost as part of community cooperation, which also increased local awareness of the project.

The modules are quite expensive compared to the initial costs of some other coral restoration techniques. However, maintenance after initial installation was minimal compared

to the costs of monitoring and maintaining coral transplants. It was about 7–10 years before the recruitment and growth of the coral community and associated organisms had transformed most of the original triangular prism modules into patches barely distinguishable from the natural reef. Thus the method would not be appropriate for those seeking "quick-fixes".

Diver over the artificial reefs 12 years after deployment. Triangular prism modules are now hard to distinguish from the natural reef (N. Thongtham).

References

Chansang, H, Thongtham, N., Satapoomin, U., Phongsuwan, N., Panchaiyaphum, P. and Mantachitra, V. (2008) Reef rehabilitation at Maiton Island: the prototype of rehabilitation by using artificial substrate in Thailand. ICRS Abstract 24-32. p. 222

Phongsuwan, N. (1991) Recolonization of a coral reef damaged by a storm on Phuket Island, Thailand. *Phuket Marine Biological Center Research Bulletin*, 56, 75-83.

Thongtham, N. and Chansang, H. (1999) Influence of surface complexity on coral recruitment at Maiton Island, Phuket, Thailand. *Phuket Marine Biological Center Special Publication*, 20, 93-100.

Thongtham, N. and Chansang, H. (2009) Transplantation of *Porites lutea* to rehabilitate degraded coral reef at Maiton Island, Phuket, Thailand. *Proceedings of the 11th International Coral Reef Symposium*, 1271-1274.

Case study provided by: **Hansa Chansang** and **Nalinee Thongtham**, Phuket Marine Biological Center, PO Box 60, Phuket 83000, Thailand; Tel: +66 76 391128; Fax: +66 76 391127; E-mail: nalineeth@hotmail.com

Case study 5

Transplantation of nursery reared corals to a degraded reef at Eilat, Israel.

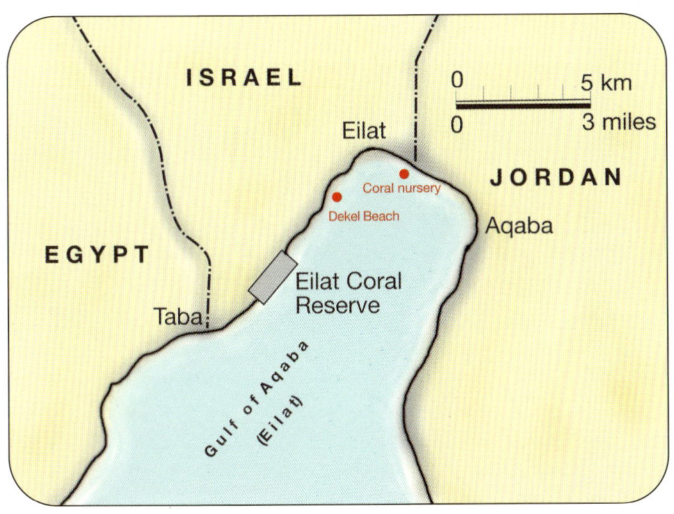

Eilat, Israel, Red Sea (November 2005 - ongoing).

Background

This pilot project aims to rehabilitate reefs at Eilat degraded through human activities including coastal development, port activities, pollution and recreational SCUBA diving, by adding nursery-grown coral colonies to degraded coral knolls at Dekel Beach to create topographic complexity and new ecological niches for marine invertebrates and fish. This project was carried out in association with the Israel Nature & National Parks Protection Authority.

Objectives

To assess 1) the survival and acclimation of nursery-grown coral colonies transplanted to a degraded reef environment, 2) their impact in terms of attracting invertebrates and fish, and 3) the ability of the transplants to contribute to the larval pool at the restoration site.

Methods

Coral fragments (1–2 cm in size) were reared for 7 months to 1.5 years at about 6 m depth in a mid-water floating nursery (see Chapter 4) where they were largely protected from coral predators and not subject to human disturbance (e.g. by snorkellers and divers). The nursery environment also differed from the natural reef in that it was sited near fish mariculture cages and thus had elevated nutrients compared to the reef, which increased coral growth rates. Prior to transplantation the corals on their plastic pin substrates were cleaned to remove algae and other sessile invertebrates and had any corallivorous gastropods removed.

880 nursery-grown coral colonies were transplanted onto 5 large knolls (totalling c. 80 m²), 7–13 m deep, 2.7 km from the nursery. The transplants were transferred in 20 minutes from the nursery to the rehabilitation site by boat, submerged in large containers filled with seawater. The plastic pins on which the colonies were raised in the nursery were inserted into holes pre-drilled in the natural substrate using pneumatic drills powered by diving tanks. A small amount of epoxy was placed in the bottom of the holes to secure the pins.

550 colonies of two locally common branching species (*Stylophora pistillata* and *Pocillopora damicornis*) were transplanted in November 2005. A further 330 colonies of six branching species (*S. pistillata, P. damicornis, Acropora variabilis, A. humilis, A. pharaonis, A. valida*), one massive species (*Favia favus*) and one hydrozoan (*Millepora dichotoma*) were transplanted in May 2007. For the second transplantation, plastic wall-plugs (see Chapter 4) were also tested as an alternative substrate for some of the transplants. For the initial transplantation, colonies were 6–10 cm in diameter and spaced c. 20 cm apart on the knolls; for the second one, they ranged from 6 cm up to 40 cm diameter (for some *Acropora*) and were spaced about 10 cm apart.

Using a compressed air drill to make holes in the degraded reef into which plastic pins or wall-plug substrates on which the coral transplants have been grown can be slotted (Y. Horoszowski).

Monitoring

A detailed analysis of survivorship, detachment, bleaching and the general health of transplanted colonies was made during the first four months after transplantation, in order to see whether the transplantation procedures had stressed the corals and whether coral colonies that had been reared in a nursery could cope with conditions on the natural reef.

Subsequently monitoring was undertaken monthly during the first year, and thereafter every 2 or 3 months. Data were collected on survivorship of transplants and controls (coral colonies growing naturally at the transplanted reef area), rates of detachment, growth, bleaching and the general health of the transplanted/control colonies; invertebrates residing in the colonies and new recruits. Planulae released from *Stylophora* transplants and controls were counted using planulae-collector nets placed on the colonies from sunset till sunrise. Data were also collected on fish abundance at the transplantation site for 1.5 years after the initial transplantation.

Ecological outcomes

New transplants had a very low mortality rate – less than 5% – after four months, indicating that colonies initially adapted well to the new environmental conditions. Survivorship was similar to that of natural colonies at the same site (the controls). Some transplants were attacked by fish: butterflyfish grazed on polyps, parrotfish damaged colonies by biting off whole branches, but most damaged colonies survived and regenerated lost parts. After almost two years, overall survival of the first transplantation was approximately 62%, slightly lower than that of established naturally-growing control colonies (73%) on the reef (equating to ~5% greater mortality per year).

The new ecological and spatial niches created by the transplanted colonies were immediately colonized by obligate coral commensal invertebrates including *Trapezia* crabs, *Spirobranchus* worms and *Alpheus* shrimps. The transplants were also found to be hosting new *Lithophaga* boring bivalves.

There was an increase in fish abundance on the transplanted knolls compared to adjacent control knolls (without transplants) where no change was observed, however, fish species richness did not increase. The transplanted nursery-grown *Stylophora pistillata* colonies were found to be releasing planula larvae during each reproductive season since they were transplanted.

Social aspects

The project did not aim specifically to involve local communities but the translocation site is in front of a diving centre and generated much interest. This led to discussions with divers and other local people and an opportunity to raise awareness of the contribution of recreational activities such as diving and reef walking to reef degradation. The diving centre has now incorporated the topic of coral degradation in diving courses, resulting in more aware divers and diving instructors. Many people are following the project's progress and show interest in the monitoring results.

Collecting planula larvae from Stylophora pistillata transplants. These have produced larvae each season after outplanting from the coral nursery (Y. Horoszowski).

One of the denuded coral knolls, 1.5 years after transplantation (Y. Horoszowski).

Resources required to rehabilitate c. 80 m² of denuded reef

Human: At least 15 people (researchers and volunteers) were involved in preparation of the nursery-reared coral colonies for transplantation (cleaning of the plastic pin substrates and removal of corallivorous snails) and the transplantation itself. Only brief training needed to be given to the volunteers who assisted with preparation of the nursery-reared corals and helped with the transplantation.

Financial resources: No data provided.

Time: Preparation of the 550 corals in the nursery for the first transplantation took 13 volunteers about one week and the transfer of the corals to the rehabilitation site and their attachment took 5 volunteers two weeks. For the second transplantation, preparation of the 330 colonies was done by 2 people over three weeks, whereas the transplantation was conducted by 8 people over 3 days. About 10–15 colonies could be prepared per hour per person and about 30 colonies could be transplanted (included drilling, epoxy mixing, tagging). Monitoring of the first transplantation took about one week per survey; this increased to two weeks after the second transplantation.

Lessons learnt

Using corals reared from small fragments in nurseries for transplantation rather than corals taken from the wild allows potentially much larger-scale restoration since large stocks of new corals can be generated with minimal impact on natural reefs.

Although we were worried that the nursery-reared corals might not survive well when transplanted to the less favourable reef environment, they showed good short and long-term survival despite some initial grazing by fishes, suggesting that nursery-reared colonies are suitable for large-scale restoration.

Other studies that have used epoxy to attach corals generally suggest that about 5–6 colonies are deployed per person-hour. However, using colonies reared on plastic pins we were able to transplant at least 30 colonies per hour. This suggests that growing corals on substrates that can be directly inserted into the reef increases the efficiency of transplantation by 5–6 times over attaching corals using epoxy putty.

Fish predation and diving activity at the site led to detachment of colonies and as part of the second transplantation wall-plugs were tested as alternative substrates to the plastic pins. Initial results showed that the wall-plugs reduced detachment losses from 10–20% (depending on species) to <5%.

References

Rinkevich, B. (2006) The coral gardening concept and the use of underwater nurseries; lesson learned from silvics and silviculture. Pp. 291-301 in Precht, W.F. (ed.) *Coral Reef Restoration Handbook.* CRC Press, Boca Raton, Florida.

Shafir, S., van Rijn, J., Rinkevich, B. (2006). A mid-water coral nursery. *Proceedings of the 10th International Coral Reef Symposium,* 1674-1679.

Case study provided by: **Yael Horoszowski,** Israel Oceanographic and Limnological Research, Tel Shikmona POB 8030, Haifa, Israel; Tel: +972 (0)4 8565273; Fax: +972 (0)4 8511911; E-mail: yaelh@ocean.org.il

Case study 6

Re-attachment and monitoring of broken fragments of *Acropora palmata* following a ship grounding in Puerto Rico.

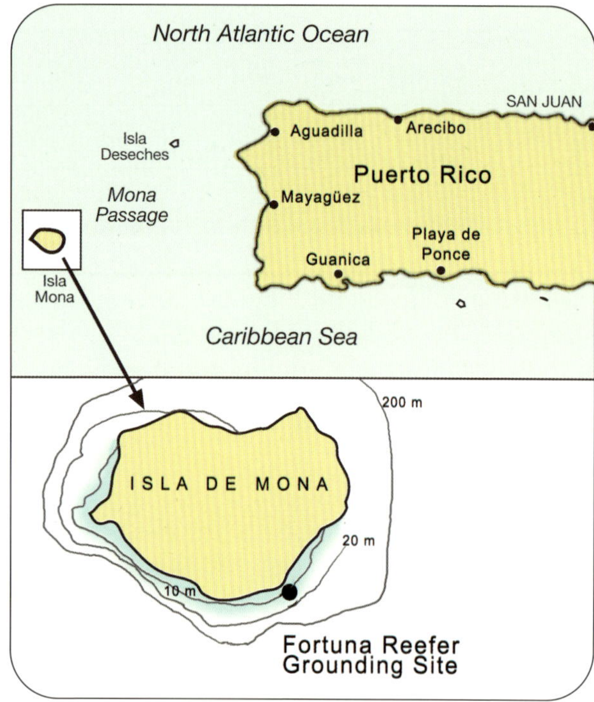

South-east coast of Mona Island, Puerto Rico, USA (1997-2008).

Objectives

To re-attach fragments of *A. palmata* (Elkhorn coral), a species listed on the US Endangered Species Act, to recreate the reef habitat and its structural relief that had been damaged by the grounding, and to reduce the mortality of broken coral fragments.

Methods

Restoration was undertaken 3 months after the grounding, during September and October 1997. A total of 1857 fragments of *A. palmata* were collected on the grounding site in depths of 1–6 m and varying in length from 15–340 cm. Some were attached to stainless steel nails that were epoxied into holes drilled into the relict reef substrate using stainless steel wire wrapped over fragments and around the nails. Others were attached to dead standing *A. palmata* skeletons using stainless steel wire. In a few cases, plastic cable ties were used but these loosened quickly in heavy surge so this technique was abandoned in favour of wire. SCUBA and free diving were used, and lift bags were employed to reposition heavy fragments. No fragments were removed from the water. Due to considerable corrosion and breakage of the original wires after about 3 years, surviving fragments were further stabilized in July 2000 using a more durable copper-nickel alloy wire (Monel 400) with, in some cases, Portland cement, particularly where fragments were located shallow, in heavy surge.

Monitoring

The fragments were monitored 1–3 times a year for the first 6 years, and then annually until the tenth year. The following data were collected: fragments present or missing; maximum length to nearest cm; orientation of attachment (up or down, with respect to orientation prior to grounding); location of attachment (relict reef substrate or dead standing *A. palmata* skeletons); condition (live or dead). Fragment

Background

The grounding of the 99-m freighter M/V *Fortuna Reefer* on 24 July 1997 and its subsequent removal impacted 2.75 ha of shallow *Acropora palmata* forereef habitat. There was total coral destruction along an area up to 30 m wide, that extended from the reef crest about 300 m seaward at 1–4 m depth. Colony breakage occurred over a much larger area, to about 7 m depth, in part due to the steel cables used to extract the vessel that dragged across the reef. Entire colonies of *A. palmata*, many of which were several metres in diameter, were crushed or dislodged and fractured by the ship, and in addition the cable sheared off hundreds of branches. Restoration was undertaken as part of a Natural Resource Damage Assessment (NRDA) settlement with the party responsible for the ship grounding. Under the Oil Pollution Act, the US government is responsible for restoring trust resources and compensating the public for lost use of natural resources; the National Oceanic and Atmospheric Administration (NOAA) was therefore able to pursue damages due to the threat of release of oil posed by the ship grounding, although no oil was spilt.

condition was assessed through estimates of tissue loss, and causes of mortality were identified as named diseases, predation (*Coralliophila abbreviata* snails or parrotfish), overgrowth by sponge (*Cliona*) or algae, or other factors. Coral growth over the wire, fusion with the substrate, and the amount of new growth was also noted. Fish abundance and species composition within the grounding site and in surrounding areas was also monitored using 15–20 30-m belt transects per survey.

Ecological outcomes

In general, there were high rates of early mortality due to wire breakage and removal of fragments during winter storms, overgrowth by bioeroding (*Cliona*) sponges, disease and predation by gastropods (*Coralliophila*). After 2 years, 57% of the fragments had survived, 26% were dead, and 17% had become detached and had disappeared from the site. The largest number of fragments died from *Cliona*, primarily because fragments were attached directly on top of this sponge.

After 3 years (i.e. by 2000), a further 8.3% of fragments had disappeared from the site, making a total of 25% loss of reattached fragments as a result of wire corrosion and breakage. The wire was also a significant cause of partial mortality, as the high surge at the shallow site loosened it and then abraded coral tissue that was in contact with it. However, there were some instances where the tissue overgrew the wire where it remained tight. Algae or *Millepora* also overgrew the wire, dividing tissue into smaller patches that slowly died. The fragments exhibited a limited ability to fuse to the substrate and only about 17% of the survivors at 3 years showed tissue growth onto the substrate.

At 5 years, coral tissue was overgrowing the new wire used to stabilize the fragments in 2000, and there was very little initial breakage. However, after a further 4 years (9 years after the initial restoration) the new wire began to break and fragments detached, partly due to numerous storms, and partly because the dead skeletons to which many larger fragments had been attached were collapsing from bio-erosion and the weight of growth of the reattached fragments.

After 10 years (2008), just under 6% (104) of the fragments were still alive, although only a small proportion of these were securely fused to their attachment sites; 26% had become detached or were missing, and 68% had died in place. About half of the survivors resembled adult colonies with tissue covering the upper skeletal surfaces, extensive branching (mean = 5 branches, 89 cm in length), and a substantial increase in height (mean 39 cm tall). The highest survival was in fragments of 20–80 cm length attached to the relict reef substrate. Overgrowth by *Cliona* was one of the most significant stressors, with about 22% of the fragments dying due to this. Ongoing sources of mortality include sponge overgrowth (6%), snail predation (8%), and disease (6%).

Acropora palmata stand adjacent to the Fortuna Reefer grounding site in 1997 (A. Bruckner).

Fortuna Reefer site one week after the grounding in 1997 showing where keel struck the reef (J. Morlock).

Social aspects

The restoration work led to heightened awareness of the importance of coral reef resources and contributed ultimately to protection of Mona Island, which was an important destination for spear-fishermen and boaters. The involvement of the Puerto Rico Department of Natural and Environmental Resources (PRDNER) and the concomitant decline of fishery resources around the island led to changes in policy and designation of most of the shallow waters around the Island as a protected area. Mooring buoys were installed in areas of high recreational use to reduce the use of anchors.

Resources required

Human resources: The initial assessment of damage was undertaken by a team of experts from the PRDNER and the NOAA Damage Assessment, Remediation and Restoration Program (DARRP) with University of Puerto Rico (UPR) professors. The restoration work was undertaken by a team of 19 marine engineers and biologists. Federal staff provided the oversight and UPR staff and students and Federal staff undertook the monitoring. Volunteers were provided by the Center for Field Studies/Earthwatch for 3 years to do one to two missions annually with 3–4 days assisting in surveys at the *Fortuna Reefer* site. They were trained by the lead scientist, and their main jobs included tagging and measuring corals, and running out transect tapes and rope to grid the site.

Financial resources: The settlement for the grounding totaled US$ 1.25 million. Of this, US$650,000 was used for the immediate restoration, and US$100,000 went to both PRDNER and to NOAA to defray assessment costs accrued during the response. An additional US$400,000 was provided to PRDNER for compensatory restoration. The Center for Field Studies/Earthwatch provided approximately US$2,000–5,000 per year for monitoring.

Time: The initial restoration in 1997 was carried out over 3 weeks. Each monitoring survey took 3–4 days using two experts to assess corals and 6–8 volunteers to assist in a range of survey tasks (photographing corals, laying out lines and transect tapes, holding measurement bars, etc.).

Lessons learnt

Despite the high cost of this project, it was felt that without intervention, a very high percentage of fragments would have died due to sand scouring, or would have been removed from the site during high wave action. After 6 years the survivorship was comparable to or higher than that on reefs following other catastrophes such as hurricanes. However, within 10 years nearly 95% of the re-attached fragments died or disappeared and only about half of the remaining fragments were in good health and resembled adult colonies. Due to the lack of studies following the fates of natural fragments over 10 years, it is not known how their fates would compare. However, areas surrounding the grounding site declined to a similar extent and there are few noticeable differences between a natural undamaged site and the restoration site at this location. Wire alone should be avoided wherever possible as a re-attachment method, because of the problems of abrasion. It was used at this site because of the high wave exposure which made it difficult to attach corals with cement due to the amount of time required for the cement to harden. Wire may however be useful to temporarily hold a fragment in place until cement or epoxy hardens.

Tissue contact between a fragment and the substrate is essential. Attaching fragments to the tops of dead coral branches did not work as the fragments continued to grow upward but failed to resheet over the existing skeletons which eventually become weakened and broke. A possible solution would be to attach fragments to the bases of dead colonies, in a vertical position, as they would be more likely to fuse and resheet over the skeleton as they grow upward.

Any coral fragments attached on or near *Cliona* are likely to die, emphasizing the importance of finding substrate where this sponge is absent.

Medium sized *A. palmata* fragments (i.e. 20–80 cm) had the highest survival and growth, whereas very small fragments and large fragments, especially those taken from the older portions of a colony, were more likely to die.

Improvements to the method should include removal of pest species like snails (*Coralliophila*) during monitoring. Two removals were undertaken, at year 8 and 9, with fewer snails and less mortality associated with snails recorded during year 9, and even lower numbers during year 10. Other approaches that should be attempted include salvage of healthy portions of diseased colonies. A pilot experiment

~60 cm long fragment of elkhorn coral Acropora palmata, broken off during the grounding which has been secured to nails in the reef with stainless steel wire during the restoration. Both original wire (overgrown) and new wire added during mid-course correction are visible (A. Bruckner).

involving the removal of branch ends from corals with White Band Disease showed high survival of the detached fragments but complete mortality of the remaining part of the coral.

Acknowledgements

Monitoring of this restoration project over this many years was made possible by the countless individual contributions of partners from within NOAA, UPR (including several graduate students), Earthwatch volunteers, the captain and crew of Mona Aquatics, PRDNER, other interested colleagues and friends. The opinions and views expressed in this document are those of the authors and do not necessarily reflect those of NOAA Fisheries or the US Government.

References

Bruckner, A.W. and Bruckner, R.J. (2001) Condition of restored *Acropora palmata* fragments off Mona Island, Puerto Rico, two years after the *Fortuna Reefer* ship grounding. *Coral Reefs*, 20, 235-243.

Bruckner, A.W. and Bruckner, R.J. (2006) Chapter 14. Restoration outcomes of the *Fortuna Reefer* grounding at Mona Island, Puerto Rico. Pp. 257-269 in Precht, W. (ed). *Coral Reef Restoration Handbook.* Taylor and Francis, Boca Raton, Florida.

Bruckner, A.W. and Bruckner, R.J. (2006) Survivorship of restored *Acropora palmata* fragments over six years at the M/V *Fortuna Reefer* ship grounding site, Mona Island, Puerto Rico. *Proceedings of 10th International Coral Reef Symposium*, 1645-1650.
[Download at: www.reefbase.org/download/download.aspx?type=1&docid=12417]

Bruckner, A.W., Bruckner, R.J. and Hill, R. (2009) Improving restoration approaches for *Acropora palmata*: Lessons from the *Fortuna Reefer* grounding in Puerto Rico. *Proceedings of the 11th International Coral Reef Symposium,* 1199-1203.
[Download at: www.reefbase.org/download/download.aspx?type=1&docid=13908]

Case study provided by: **Robin J. Bruckner** and **Andrew W. Bruckner**, NOAA Fisheries, Office of Habitat Conservation, 1315 East West Highway, Silver Spring, MD 20910, USA; Tel: +1 301-713-0174 x153; Fax: +1 301-713-0184; E-mail: Robin.Bruckner@noaa.gov

Case study 7

Coral transplantation, using ceramic coral settlement devices, on reefs damaged by bleaching and *Acanthaster planci*.

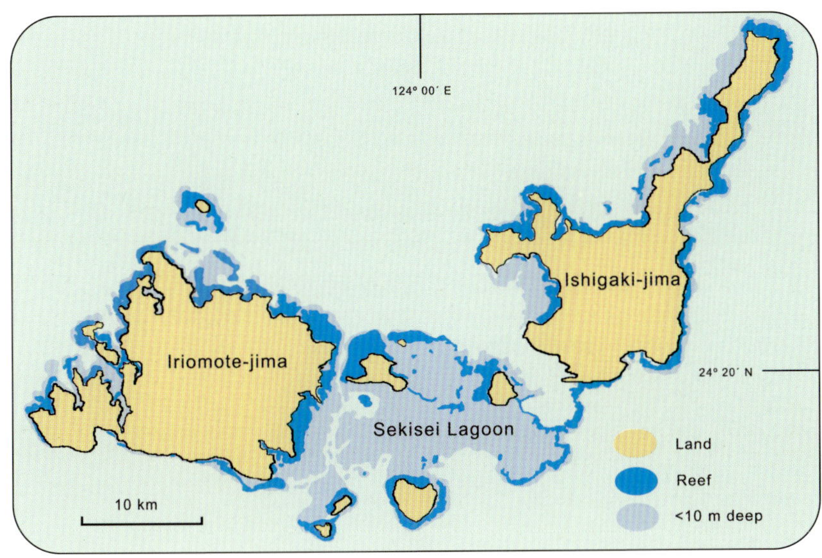

Sekisei Lagoon, Iriomote-Ishigaki National Park, South Ryukyus, Japan (2002 onwards).

Background

Following outbreaks of Crown-of-thorns starfish, *Acanthaster planci*, and then coral bleaching in 1998, reefs in Sekisei Lagoon that had high coral cover in the 1970s were severely damaged. Reefs on the northern edge of the lagoon suffered high mortality in the 1998 bleaching but prevailing currents swept larvae from the less affected interior and southern reefs to them, which enabled recolonisation. However, later bleaching events in 2001, 2003 and 2007 caused mass coral mortality in these reefs, while the northern reefs recovered. The interior and southern reefs showed little recovery due to low recruitment, a result of the local currents sweeping coral larvae produced on the northern reefs away from the lagoon. *Acanthaster* outbreaks still occur, but the coral-eating starfish are continually removed under a government project. Since 2002, custom-made ceramic coral settlement devices (CSD), which are designed 1) to encourage settlement of natural coral larvae, 2) to enhance their survival once settled, and 3) to be easily handled underwater and transplanted, have been tested as a means of accelerating the recovery of the Sekisei reefs.

Objective

Recovery of the reefs to their condition in the 1970s when the first large *Acanthaster* outbreak occurred.

Methods

Prior to restoration, reefs were surveyed and mapped using aerial photography, currents were simulated and recruitment rates were studied (using recruitment tiles and surveys of juveniles). Sites at depths of 4–10 m, with low coral cover, poor recruitment, and no sedimentation were selected for transplantation. CSDs were used both to collect larvae and to transplant corals.

In 2004–2005 about 81,000 ceramic CSDs were deployed at 15 sites in the lagoon where larval supply was predicted to be plentiful. The CSDs were deployed on the bottom at less than 6 m depth stacked in stainless steel frames (each holding around 700) shortly before the known date of mass coral spawning, which is around the first full moon after sea temperatures reach 26°C in Sekisei Lagoon (May). About 31,000 CSDs were retrieved in 2006 along with any juvenile corals that had settled and grown on them. They were kept immersed in seawater and transported by boat to the transplantation site (a distance of c. 18 km) where they were sorted and those with juvenile corals (about 20%) selected for transplantation. SCUBA divers inserted the leg of each of these CSDs into a pre-drilled hole in the substrate and secured it with epoxy. For hard substrate, CSDs with 1-cm legs were used, whereas for stable rubble or areas of dead branching coral, CSDs with 13-cm legs were used. Most of the transplants on the CSDs were

species of *Acropora*, which dominated the mass spawning.
The method has been repeated in subsequent years.

Three-year old recruits that have settled on ceramic coral settlement devices (CSDs). Top: Millepora transplanted on CSD with 13-cm leg wedged in dead branching coral (S. Fujiwara). Above: Acropora transplanted on CSD with 1-cm leg fixed to coral rock (S. Fujiwara).

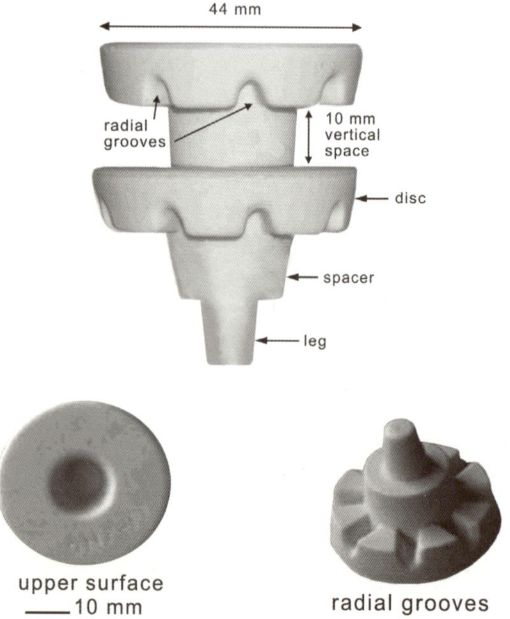

Top: Two CSDs stacked together (based on Okamoto et al., 2008). Above: Coral settlement devices (CSDs) to show upper surface and radial grooves.

Monitoring

10% of the transplanted CSDs were marked with plastic tags and monitored after 1, 3, 6 and 12 months during the first year (2006) and then every 6 months for the following four years. Data collected included coral size and genus, dead tissue area per colony, bleaching, breakage, predation, algal overgrowth and sedimentation.

Ecological outcomes

5400 CSDs with attached corals were transplanted in February 2006, a further 805 in December 2006, and 1271 in January 2008. CSDs were evenly distributed at a density of 10/m². Average survival rate of the first batch in August 2006 (6 months after transplantation) was 78.5 %, and corals had doubled in size. By February 2007, survival had decreased to about 40% due to disturbance by a typhoon in September 2006, but corals had increased in size by three times. By February 2008, survival rate was about 30% due to severe bleaching in August 2007, and due to partial mortality the average coral size had halved.

Social aspects

The project was initiated by the national government but involved local people, fishermen and the municipality, who are supportive because of the important role of the reef in the park, fishery, and for tourism. The project was part of a larger restoration programme developed at a workshop attended by members of government, the municipality and the local community, including fishermen.

Resources required to collect coral larvae on CSDs and transplant corals on CSDs to c. 750 m²

Human resources: A scientific committee of 7 people advised the project; 4 government staff supervised it. 3 staff managed the project and there were 6 divers. Fishermen participated as divers and deployed the CSDs and later transplanted them; there was no training involved as divers were already sufficiently experienced to handle the coral colonies on the CSDs.

Financial resources: The pilot study was financed by the Government and costs are not known. Following the pilot's success, Japan's environment ministry has now devoted a budget of about US$ 430,000 a year to the restoration of the Sekisei Lagoon.

Time: Project design: 2 years; collection of coral larvae on CSDs: 1.5 years; transportation to damaged reefs:1–2 hours; transplantation: 1 day; monitoring: 2 days for each survey.

Lessons learnt

The financial support from the Government and technical advice from the scientific committee were essential for the success of the project.

Annual fluctuations in the success of natural larval settlement on the CSDs is a problem and so trials are underway using larvae collected from spawning slicks that are cultured in tanks.

The fact that the CSDs are raised off the seabed helps to prevent damage to and loss of juvenile corals due to drifting rubble and sand, particularly during typhoons.

References

Fujiwara, S. and Katsukoshi, K. (2007) Coral reef restoration using larvae collector. *Proceedings of ITMEMS 3, Global Problems, Local Solutions, Cozumel, Mexico, 16-20 October 2006 (online).*

[Powerpoint (31 slides) downloadable from: www.itmems. org/itmems3/NEW 02 DISASTER MANAGEMENT/05 T2 PRESENTATIONS/07_ Fujiwara.ppt]

Normile, D. (2009) Bringing coral reefs back from the living dead. *Science*, 325, 559-561.

Okamoto, M., Nojima, S., Fujiwara, S. and Furushima, Y. (2008) Development of ceramic settlement devices for coral reef restoration using in situ sexual reproduction of corals. *Fisheries Science*, 74, 1245-1253.

Okamoto, M. and Nojima, S. (2004) Chapter 6. Development of underwater techniques for coral reef restoration. Pp. 50-61 in Omori, M. and Fujiwara, S. (eds). *Manual for restoration and remediation of coral reefs*. Nature Conservation Bureau, Ministry of Environment, Japan.

[Download available at: www.coremoc.go.jp/report/RSTR/RSTR2004a.pdf]

Case study provided by: **Shuichi Fujiwara**, IDEA Consultants Inc., Naha, Okinawa 900-003, Japan; Tel: +81-98-868-88 84; Fax: +81-98-863-7672; Email: fjw20240@ideacon.co.jp

Case study 8

Transplantation of corals to a traditional no-fishing area affected by coral bleaching in Fiji.

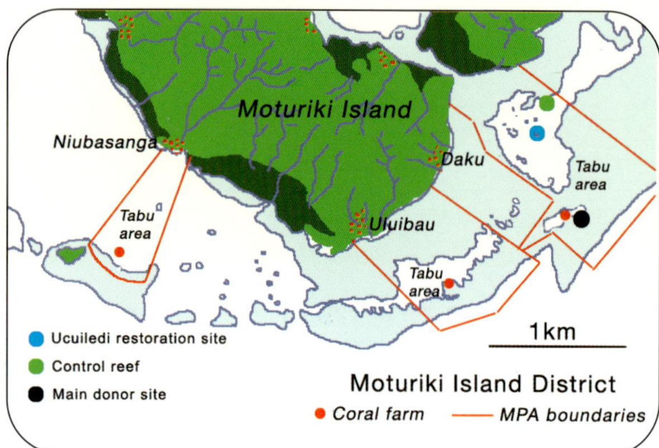

Ucuiledi Reef, Moturiki Island, Fiji (August 2005–May 2006).

Background

Reefs within a traditional no-fishing or *tabu* area were damaged by bleaching events in 2000 and 2002. The *tabu* area is managed through a community-based process with help from Partners in Community Development Fiji (PCDF), and partial support from the CRISP program.

Objective

Restoration was undertaken at the request of the local communities in order to help restore fish populations and thus improve food security and community livelihoods. The project was also testing a low-cost restoration method for use in shallow low-energy reef areas, with an emphasis on local community involvement.

Methods

Coral colonies and fragments (20–50 cm diameter) were chosen by swimming over reef areas adjacent to the *tabu* area, and collecting those that were already broken or unlikely to survive (e.g. those very close to the surface, overgrown colonies, fragments lying on the sandy substratum that had been detached by anchors or divers' fins, etc.). Collection was done manually, without the use of tools, using free-diving at 1–4m depth. In addition, about a hundred 1–2 year old *Acropora* colonies (c. 10 cm diameter) were obtained from a coral farm on nearby Cagalei Island. Corals were transported by boat, exposed to air for 30–60 minutes but regularly sprinkled with fresh seawater from a bucket.

80% of transplants were *Acropora* (particularly *A. muricata*). The rest were mainly *Pocillopora*, *Stylophora* and branching and massive *Porites*.

Corals were transplanted at a density of 1 colony or fragment/m², and were not placed near natural colonies. Three planting techniques were tested:
- Placing large colonies directly onto rubble, and stabilizing them with rocks.
- Inserting fragments into holes and crevices ("plug-in" method) in the hard substratum – used for the majority of specimens.
- Cementing farmed coral colonies to the substrate using regular cement.

Maintenance after transplantation included:
- Replanting loose coral fragments that had been moved by waves into a position where they were likely to die,
- Removing and destroying or relocating predators, including crown of thorns starfish *Acanthaster planci*, *Drupella* snails, and *Culcita* sea stars,
- Cleaning plots and transplanted corals of any rubbish or loose seaweed,
- Maintaining equipment used for monitoring (marker posts, ropes, etc.).

Monitoring

Although planned for 18 months, monitoring was undertaken for only 9 months as most of the corals died. A first visit was made at one month to:
- Identify mortality associated with initial translocation of the coral transplants,
- Identify potential methodological weaknesses, and changes needed to the transplanting methods, and
- Ensure that monitoring data collection methods were clearly understood and agreed by scientists and field assistants.

Subsequently monitoring was carried out at 3, 6 and 9 months with about one third of the total restored area being surveyed during each survey. Data were collected on transplant survival, coral cover, fish and benthic organism colonization, and natural coral recruitment, using the following methods on 12 interspersed plots (3 restored and 9 control plots) at the rehabilitation site and on 3 plots at a control site (a similar situated nearby reef in the same lagoon).

- Line-intercept transects to assess substrate composition and cover of sessile benthic community (life-form categories for corals).
- 2-m wide belt transects along the same permanent transects to assess transplant mortality and partial mortality, colony attachment to the substrate, incidence of bleaching or disease, and presence of predators.
- Visual censuses of fish in the belt transects. (The same two people carried out all the fish surveys to reduce sampling error.)
- Counts of invertebrates on 3 restoration and 3 non-restoration plots at the rehabilitation site and on the 3 plots at the control site.

Ecological outcomes

The rehabilitation covered about 2000 m² within a 1 ha patch reef, and about 2000 coral colonies and fragments were transplanted in total. After one month, the mortality rate was only 0.6% and over 80% of the transplanted colonies were in good health (defined as having less than 5% of their living tissue dead) and branching species showed growth onto the substrate. After 6 months, only 1.1% of transplanted colonies had died and surviving colonies were still in good health. At this stage, 95% of the farmed colonies had self-attached by tissue expansion over the cement; for the plug-in method, there was 62% self-attachment; but for the placed-on method, only 33% of transplants had firmly attached. At the 1, 3 and 6-month

Farmed coral colonies reared for one year from fragments at the coral nursery in the Cagalei Island tabu area (S. Job).

Transplantation of a farmed Acropora colony (grown from a fragment for one year in a nursery) by a local free diver from Ucuiledi village who volunteered to help (S. Job).

surveys, 12–16% of colonies showed between 6% and 50% partial mortality. This included the dead parts of the fragments and colonies already present (which were not removed prior to transplanting).

However, by 9 months, 75% of the transplants were completely dead and about 20% were severely damaged due to a late bleaching event in May. The remaining 5% that were still alive in June 2006 showed varying degrees of bleaching, mostly on the upper parts. Many naturally occurring colonies of Acropora and other genera were also observed bleached at this time on the restoration reef, the control reef, the donor reef and Cagalei Island indicating that the bleaching was unrelated to the transplantation. However, there was less mortality of corals at the donor site in the outer lagoon. Warmest sea temperatures are normally in January–April in Fiji.

Social aspects

Local communities who owned the tabu area where the restoration took place assisted with the transplantation and the project raised community awareness of the need for reef conservation.

Resources required to transplant c. 2000 m²

Human resources: 4 persons: 2 marine biologists, 1 community facilitator (who also helped in the field) and 1 boat driver. The project coordinators supervised and advised the team doing the restoration work. Monitoring at 1, 3 and 6 months was done by a Foundation of the Peoples of the South Pacific International (FSPI)/PCDF team, and the final monitoring by a team comprising the same staff from FSPI/PCDF/SPI Infra, but with new staff from Institute of Applied Sciences (University of the South Pacific) and a private consultant.

Financial resources: US$12,000 (FSPI).

Time: Project design: 1 week; coral collection/transportation/transplantation: 1 week; baseline survey: 4 days; monitoring: 2 days/survey.

Lessons learnt

Despite the relatively harsh conditions in which the corals were transported (Methods), the results over the first 6 months suggested that the transplanted colonies recovered well from the stress of being transplanted. Thus, where time and budgets are limited, simple methods can be successful.

Donor and transplant sites should be as similar as possible with respect to environmental conditions (wave, current, depth, temperature, light, and disturbance regimes). Although, corals sourced from the outer lagoon and transplanted to the inner lagoon reef survived well initially, in the longer term they seemed poorly adapted to the more extreme conditions experienced there.

The predominantly branching *Acropora* spp. used as transplants were not common on patch reef tops such as the rehabilitation site. It might have been prudent to choose growth-forms and genera more suited to the mid-lagoon habitat.

Acropora muricata colony 9 months after transplantation using the 'placed-on' method. This rare survivor still shows some bleaching of branch tips (S. Job).

The 'plug-in' method was quick and easy and may be appropriate for restoring reef areas dominated by dead colonies/coral rock into which branches can be inserted, but it is restricted to small branching corals. It is important to choose appropriately-sized holes so that corals are held securely in place and to ensure that living tissue is in direct contact with the substrate to maximize self-attachment. Where holes are too large, fragments can be wedged in place with coral rubble.

The 'placed-on' method resulted in only 33% self-attachment after 6 months and is only appropriate for low-energy environments where the weight of the branching colony or large fragment is sufficient to keep the transplant stable until it can self-attach or its base can settle into sand. Where possible, such transplants should be positioned where they will be relatively sheltered and wedged with rocks if necessary.

The cement method resulted in 95% self-attachment after 6 months and was suitable for corals that could not be easily plugged into holes and that were too small and light to be placed directly on the substrate (small to medium sized rounded colonies, massive colonies, and farmed corals grown on cement discs).

Monitoring should be undertaken for at least one full year to take account of seasonal changes in the environment at the transplant site, and to determine whether transplants can survive during the worst conditions.

References

Job, S., Bowden-Kerby, A., Govan, H., Khan. Z. and Nainoca, F. (2005) *Field Work Report on Restoration Work, Moturiki District, Fiji Islands.* SPI INFRA/FSPI. 31 pp.

Job, S., Fisk, D., Bowden-Kerby, A., Govan, H., Khan. Z. and Nainoca, F. (2006) *Monitoring report on restoration work: 1, 3, 6, and 9 month surveys. Moturiki District, Fiji Islands.* Coral Reef Initiative for the South Pacific. SPI INFRA/FSPI/PCDF. 33 pp.

[Download at: www.crisponline.net/Portals/1/PDF/C2B1_Motoriki_report.pdf]

Case study provided by: **Sandrine Job**, SOPRONER-GINGER, Imm. «Oregon», 1 rue de la République, BP 3583 - 98 846 Nouméa Cedex, New-Caledonia. Tel : +687 28 34 80 ; Fax: +687 28 83 44; E-mail: sandrinejob@yahoo.fr

Case study 9

Transplantation of coral fragments onto artificial reefs at a hurricane-damaged site in Cozumel, Mexico.

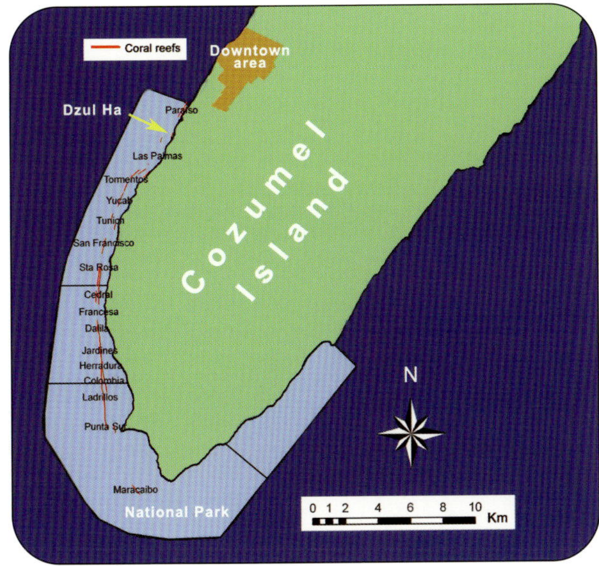

Dzul Ha Reef, inside the Parque Nacional Arrecifes de Cozumel, Cozumel Island, Quintana Roo, México (May 2007 onwards).

Background

Hurricanes Emily (July 2005) and particularly Wilma (October 2005) caused widespread damage to the reefs of Cozumel. The shallowest reefs, such as Dzul Ha, were among the most severely damaged. Cruise Lines International Association (CLIA), Conservation International (CI) and a local business provided financial support to install Reef Balls™ to create an artificial reef, guided by managers from the Marine Park.

Objectives

The aims of the project were a) to reduce tourist pressure on natural reefs by providing an alternative site for snorkelling and diving, b) to establish a demonstration restoration project, and c) to promote the recovery of endangered coral species (notably the Elkhorn coral *Acropora palmata* and the gorgonian *Plexaura homomalla*, both of which are protected species in Mexico). A special effort was made to re-establish *A. palmata* as only a few colonies survived in Cozumel after the hurricanes (and only a single colony survived at Dzul Ha). The artificial reef is also intended to encourage Marine Park staff to undertake monitoring to determine trends in reef health.

Methods

Fragments of five species of hard coral (*Agaricia agaricites*, *A. tenuifolia*, *Porites porites*, *P. astreoides* and *Siderastrea siderea*) and three soft corals (*Eunicea* sp., *Pseudoplexaura* sp. and *Plexaura homomalla*) were collected on site from live colonies detached by hurricane action and were transported in plastic buckets filled with fresh seawater to a work table (see www.reefball.com/reefballcoalition/reefballattachementsystem.htm for methodology). Fragments of the locally rarest species, *Acropora palmata*, were taken from a site located 4 km away (10 minutes by boat). Before transplantation at the site, fragments were fixed in cement plugs and placed in nurseries located close to the work table at 1 m depth to allow them to acclimate for a few hours before transplantation.

Initially, 32 Reef Balls were deployed in May 2007. Support from a local business, the Marine Park, the Reef Ball Foundation and Reef Ball Mexico, permitted a second phase of the project and by June 2007, a total of 73 Reef Balls had been deployed at 1.5–5.5 m depth over an area of about 18,500 m². Between one and seven fragments were attached to 46 of the Reef Balls using cement and epoxy putty. In total 81 coral fragments and 7 adult colonies were transplanted. The cement plugs created a solid base for the fragments, and the epoxy putty was used to attach these to the Reef Balls, although in some cases, the coral fragment or colony was attached directly to the Reef Ball using epoxy putty.

Monitoring

Monitoring by Marine Park staff started in June 2007, with surveys every 2 months, and is on-going. Coral transplant survival, tissue growth, bleaching, mortality (algae overgrowth, predation, etc.) and new recruits of hard coral are recorded on each Reef Ball. Adult and juvenile fish abundance on Dzul Ha reef and fish colonization of the Reef Balls are recorded using 2-m wide, band-transects and the "point count" visual census method, respectively.

Ecological outcomes

By June 2008 (one year after deployment), 73% of the fragments on the Reef Balls had survived and showed good growth. However, 25% of the fragments were overgrown by filamentous and blue-green algae. At the same time, recruits of *Porites* sp., *Agaricia* sp., *Favia* sp., *Siderastrea* sp., *Millepora* sp., sea rod (gorgonian) and the non-zooxanthellate coral *Stylaster roseus* were found on 65% of the Reef Balls, mainly near the base, inside or near holes. By August 2007, 32 of the transplanted *A. palmata* colonies were well-established.

Total fish density at the site increased following deployment of the Reef Balls, with parrotfishes (Scaridae), grunts (Haemulidae) and damselfishes (Pomacentridae) being the most abundant families. The increase in parrotfish abundance and the presence of the sea urchin *Diadema* is helping to control algal overgrowth. 45 fish species use the Reef Balls, the most abundant being juvenile wrasses, parrotfish and grunts seeking shelter, suggesting that the structures are functioning as "nurseries". Territorial species such as Dusky damselfish (*Stegastes adustus*) and Yellowtail damselfish (*Microspathodon chrysurus*) use the Reef Balls for feeding and shelter, and Sergeant majors (*Abudefduf saxatilis*) have been recorded depositing patches of eggs inside the structures.

Marine Park biologists continue to monitor the progress of the intervention, and in particular the impact of coral predators such as the coral-eating snails, *Coralliophila* sp.

Divers attaching coral transplants already embedded in cement plugs to a Reef Ball (M. Millet Encalada).

Social aspects

The project had additional benefits in terms of environmental education and awareness raising. High school students from Cozumel and local people visit the site. Educational signs about reefs have been put up on land. The site has been made more attractive to snorkellers by installing the Reef Balls at readily accessible depths and extending snorkelling trails.

Summer day-camps for schoolchildren are organized to teach them about marine conservation. Courses on environmental education, the rehabilitation project, and good reef practices have also been provided. Groups of 12–15 boat captains and local snorkel guides have been trained every month since April 2008, and local guides have also been trained about the marine park. A Coral Team comprising coral experts from the region, park staff and scientists, and Reef Ball Foundation trainers was formed at the Park to train park staff from both Cozumel and Cancun Marine Parks in techniques for coral rehabilitation, and they and local and international volunteers were certified to Level II Coral Propagation & Planting Specialists (a Reef Ball Foundation qualification; see www.reefball.org/volunteer.htm).

Resources required to establish 73 Reef Balls over 18,500 m² and transplant 88 corals

Human resources: Project planning (6 persons); deployment of Reef Balls (8 persons); coral transplantation team (10 persons); designing terrestrial interpretive signs (4

*Elkhorn coral (*Acropora palmata*) fragments embedded in plugs of cement in plastic cups prior to outplanting to the Reef Balls (M. Millet Encalada).*

persons); installation of buoys and terrestrial fencing
(6 persons).

Financial resources: Total budget US$ 25,000. Materials
(pliers, cement, epoxy putty, plastic cups for making
concrete plugs, work table, etc.) were provided by the Reef
Ball Foundation, and Reef Ball Mexico provided funding for
training of the coral team. The Parque Nacional Arrecifes de
Cozumel provided SCUBA equipment and boats. Reef Ball
Mexico hosted the international experts and Coral Team
volunteers.

Time: Project design: 3 months (permission, identification of
site for Reef Ball deployment and establishment of the
monitoring base line). Reef ball deployment: 1 month;
terrestrial and marine (buoys) signs: 2 months;
coral transplantation: 1 week.

Lessons learnt

This kind of project needs to be well-publicised in the media
so that local people recognize the importance of marine
conservation.

Local people need to be involved in the monitoring activities
to promote public awareness of conservation and
rehabilitation projects in marine protected areas.

Corals transplanted near the tops of the Reef Balls were
subject to damage by snorkellers, so in shallow water sites
frequented by snorkellers, corals should be transplanted on
the middle or basal parts of the structures.

Within 7 months there were over twice as many natural
coral recruits on the Reef Balls as transplants.

Coastal development near the site has had direct impacts
such as increased sedimentation on the reef; environmental
education programs are needed to develop awareness of
such issues in sectors such as construction, tourism and
local government.

*Park staff monitoring the growth of an Elkhorn coral (Acropora palmata)
transplant inserted into a "coral plug adapter" hole in a Reef Ball
(M. Millet Encalada).*

Reference

Álvarez-Filip, L. and Gil, I. (2006) Effects of Hurricanes Emily
and Wilma on coral reefs in Cozumel, Mexico. *Coral Reefs*,
25, 583.

See also: www.reefball.org/album/mexico/cozumel/National
MarineParkProjects/PlayaDzulHa/index.html and
pyucatan.conanp.gob.mx/cozumel.htm

Case study provided by: **Marinés Millet Encalada,**
Departamento de Investigación y Monitoreo, Parque
Nacional Arrecifes de Cozumel, Oficinas de la CONANP, CP
77600, Cozumel, Quintana Roo, México. Tel: +52 (987)
8724689; Fax: +52 (987) 8724275; E-mail: yectecan@
yahoo.com.mx

Case Study 10

Rehabilitation of a reef damaged by blast-fishing in the Philippines by stabilizing rubble using plastic mesh.

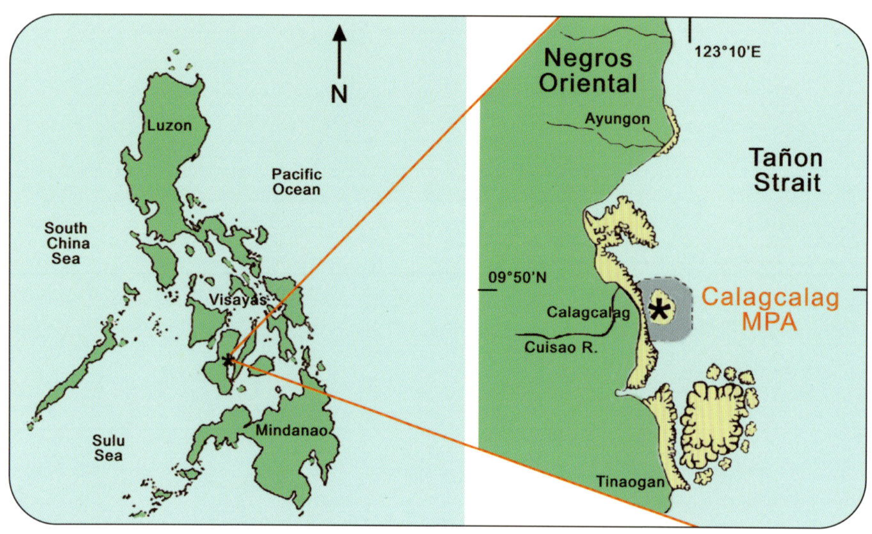

Calagcalag Marine Protected Area, Central Visayas, Philippines (June 2003–July 2005).

Background

Reefs in the 10.4 ha Calagcalag MPA (established in 1988) had been damaged by blast fishing until the mid-1980s. By 2003, the reef flat, which rises to 8 m depth, was dominated by a 2400 m² rubble field which had shown little recovery via natural recruitment over a 10 year period. While rubble covers only about 8% of the total reef area, it is located in the reef zone which would normally support the greatest coral cover and fish diversity. Despite the existence of the MPA, fishers were reporting little improvement in their catch, a breakdown of management efforts, and regular poaching within the reserve.

Objectives

The primary aims were to stabilize rubble substrate using plastic mesh in order to improve coral recruit survival and to kick-start fish habitat re-establishment through provision of rock piles transplanted with coral colonies. A secondary objective was to improve MPA management via capacity building and establishment of better relations between the managing organization and local government.

Methods

Five 17.5 m² plots were established: three in June 2003 (coral spawning season) and two in October 2003 (prior to the storm season). Locally-available plastic mesh (2-cm mesh size), was laid on the rubble and anchored with rebar stakes. Holes cut in the mesh to accommodate existing coral heads acted as additional anchorage. Hollow, pyramid-shaped rock piles (each 0.5 m² and 1 m in height) were constructed onshore by local fishers using reef rock and cement and positioned on the mesh both to hold it down and provide topographic complexity to attract fish.

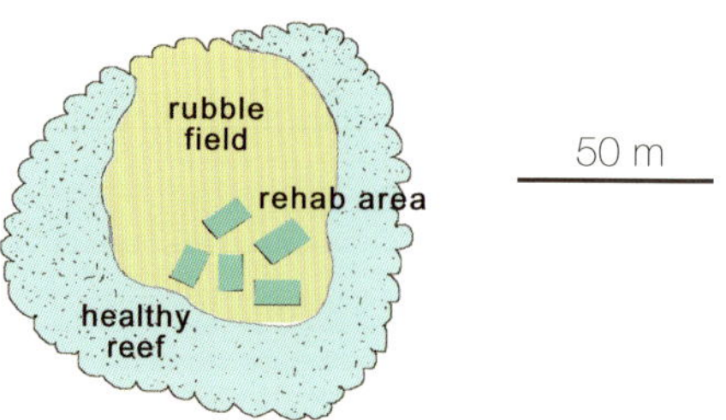

Coral transplants were obtained from the surrounding healthy reef, using corals of opportunity (natural detached fragments with a poor chance of survival), and so did not require transport. They were either attached directly to the mesh with cable-ties or fixed with cement or epoxy onto the rock piles by SCUBA divers. Approximately 75 fragments were transplanted per plot and species included *Acropora subglabra*, *Porites cylindrica*, *Pocillopora verrucosa*, *Echinopora horrida* and *Hydnophora exesa*.

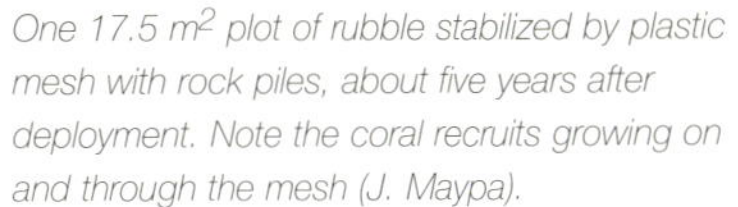

One 17.5 m² plot of rubble stabilized by plastic mesh with rock piles, about five years after deployment. Note the coral recruits growing on and through the mesh (J. Maypa).

A tagged Acropora recruit on plastic mesh, about 11 months after deployment. Note the coralline red algae colonising the mesh and stabilized rubble (L. Raymundo).

Naturally recruited coral colonies (c. 4 years old) on rock piles about five years after deployment in August 2008 (K. Rosell).

Monitoring

The site was divided into three zones for monitoring: stabilized plots, the adjacent healthy reef and the unstabilized rubble field. Each zone was monitored three to four times per year for three years after the stabilization. The site was also revisited in 2008. The numbers of each species of fish were assessed using underwater visual census for 50 m x 10 m areas in each zone and lengths of each fish estimated to ±2 cm to allow conversion to biomass. All coral recruits on the stabilized plots were counted at each survey and a subset of these were tagged in order to follow survival as were some recruits that had settled on the unstabilized rubble. Transplant survival was also recorded.

Ecological outcomes

The rehabilitation area covered 500 m², or about 20% of the 8-m deep rubble field. For the plots established in June 2003, an encrusting community composed of turf algae, diatoms and crustose coralline algae developed on the mesh within three weeks of deployment, and 1-cm diameter coral recruits began appearing by September 2003. The number of recruits continued to increase from a mean of 0.5 individuals per m² in September 2003 to 4.5 individuals per m² by March 2005. In contrast, recruits did not appear on the October-deployed plots until one year later and the number of recruits on these plots did not significantly increase over the next five months. The generic composition of the recruits (mainly Faviidae, Pocilloporidae, Poritidae and Acroporidae) generally reflected that of the surrounding healthy coral community. Coral recruits survived significantly better on the mesh net (c. 63% survival over 10 months), than on surrounding coral rubble (c. 6% survival over the same period), and recruits attached to mesh and underlying rubble began to consolidate it within one year. Mean diameter of plot recruits was over 6 cm at 10 months whereas those on the unstabilized rubble, which generally showed abrasion and partial mortality, remained at 2–4 cm. Recruitment was lower on the rock piles than on mesh/rubble, but transplant survival was higher on rock piles than on mesh.

Fish appeared to recruit to the rock piles within days and within three years there was a shift from the depauperate fish community characteristic of rubble fields to one intermediate between this and that of the surrounding healthy reef. As MPA management had its effect, fish biomass rose in all three zones with the biomass on the stabilized plots becoming similar to that on the adjacent healthy reef (over 15 kg per 500 m² after three years), whereas that on the unstabilized rubble lagged significantly and was about a third of that in the two other zones at the final survey.

There was significantly less macroalgae on the mesh than on the surrounding rubble. Algae, specifically *Padina* was seasonally highly abundant on the rubble field, but the plots stabilized by plastic mesh had almost no macroalgae. This could have been because the mesh deterred algal settlement or attachment, or because herbivory was higher within the mesh plots.

Social aspects

Calagcalag Bakhawan Fisher's Association (CABAFA), responsible for management of the MPA, participated in the restoration. Two fishers were trained in SCUBA and basic monitoring techniques. CABAFA also received a patrol boat, basic management and enforcement equipment, and a workshop on enforcement of MPA regulations. The project also helped CABAFA improve relations with the mayor, in the hope that he would continue to provide support after the scientists had left. CABAFA showed improved enthusiasm, as demonstrated in increased enforcement and willingness to participate in workshops for monitoring and enforcement.

Resources required to stabilize about 500 m² of rubble

Human resources: 3–4 researchers from Silliman University were involved in setting up the project and monitoring. A Community Organiser was the only full-time employee and was hired for the last year of the project. There was also a part-time project administrator, an assistant provided by

Fish attracted by a rock pile holding down the plastic mesh, about 3 months after deployment (L. Raymundo).

CABAFA, and research assistants from the Department of Environment and Natural Resources (DENR).

Financial resources: The cost of materials and labour for set-up averaged US$ 75 per 17.5 m² per plot. The low cost was in part due to the locally available materials and the voluntary assistance provided by CABAFA. Had the entire 2400 m² rubble field been covered with mesh, the initial outlay would have been an estimated US$ 10,560, whereas establishing rehabilitation "islands" throughout the area (at 5 plots per 500 m²) would have cost about US$ 3300. Monitoring was an additional cost.

The total budget over three years was about US$ 35,000, some of which was used for a patrol boat, signage and flashlights for enforcement. The financial breakdown was approximately as follows:-
year 1: materials – 60%, monitoring – 40% (includes per
    diems, boat/air tank rental, equipment);
year 2: materials – 40%, monitoring – 60%;
year 3: monitoring – 60%, workshops – 40%;
year 4: workshops – 30%, salary – 50%, monitoring – 20%.
A Community Organiser was funded for one year through a grant; other salaries were paid by either government or university and are additional to costs listed.

Time: The initial set-up of the mesh plots took one week and required the greatest time and financial input. Monitoring was generally possible with one full day of diving. Working with the community required a series of regular meetings, discussions and workshops, and was the main focus of the last year of work.

Lessons learnt

The methods appear cost-effective, given the increases in recruitment and the change in fish community seen within 3 years, and are probably suitable for plots up to 25 m². The plots needed no maintenance once established, in terms of additional anchoring or cleaning to remove fouling organisms.

The case study demonstrates the value of involving local communities from the project planning stage so that they understand the objectives and potential benefits to themselves of the improved conditions. The researchers felt that the project would have been more successful if a long-term relationship with the community had been maintained, even if this involved only annual monitoring visits.

Improved enforcement of the MPA was reflected in a rapid increase in target fish biomass for all three monitored zones but the results also show that improved enforcement had greatest impact on rehabilitated areas and healthy reef rather than on the rubble fields. This suggests that efforts to protect reefs with extensive rubble that show no sign of recovery will be a waste of limited resources.

Fragments cemented to the rock piles survived better than those attached with cable-ties to the mesh. This appeared mainly due to a failure of the cable-ties to prevent movement and abrasion of transplants on the mesh.

The stabilisation of rubble with mesh did not necessitate coral transplantation, as recruits were abundant and survival was high on the mesh substrate, and the site had mature coral colonies nearby which provided a supply of coral larvae. However, given the ready availability of loose fragments of branching species at the site, these were added to augment habitat complexity on the mesh areas.

A recent visit shows that 5 years after deployment, all the structures remain intact, coral recruits are up to 15–18 cm in diameter, and the community consists of 17 genera of reef-building corals. The rock piles continue to provide habitat for fish and surfaces for invertebrate recruitment but the mesh still remains visible. The critical question is whether the mesh will be completely overgrown over the next 5 years.

Reference

Raymundo, L.J., Maypa, A.P., Gomez, E.D. and Cadiz, P.L. (2007) Can dynamite-blasted reefs recover? A novel, low-tech approach to stimulating natural recovery in fish and coral populations. *Marine Pollution Bulletin*, 54, 1009-1019.

Case study provided by: **Laurie Raymundo**, Marine Laboratory, University of Guam, UOG Station, Mangilao, GU 96923; Tel: +671 735-2184; Fax: +671 734-6767; E-mail: ljraymundo@gmail.com